Ferns

and Fern Allies of the Trans-Pecos and

Adjacent Areas

Ferns

and Fern Allies of the Trans-Pecos and Adjacent Areas

Sharon C. Yarborough &
A. Michael Powell

Texas Tech University Press

This book was set in Cheltenham BT. The paper used in this book meets the minimum requirements of ANSI/NISO Z39.48-1992 (R1997). ∞

Printed in USA
Design by Brandi Price

Library of Congress Cataloging-in-Publication Data

Yarborough, Sharon C.
    Ferns and fern allies of the Trans-Pecos and adjacent areas /
Sharon C. Yarborough and A. Michael Powell.
        p. cm.
    Includes bibliographical references (p.   ).
    ISBN 0-89672-476-X cloth : alk. paper)
        1. Ferns—Trans-Pecos (Tex. and N.M.)—Identification.
2. Pteridophyta—Trans-Pecos (Tex. and N.M.)—Identification.
3. Ferns—Trans-Pecos (Tex. and N.M.)—Pictorial works.
4. Pteridophyta—Trans-Pecos (Tex. and N.M.)—Pictorial works.
I. Powell, A. Michael. II. Title.
    QK525.5.T6 Y37 2002
    587'.09764'9—dc21

                                        2001006536

02 03 04 05 06 07 08 09 10 / 9 8 7 6 5 4 3 2 1

Texas Tech University Press
Box 41037
Lubbock, Texas 79409-1037 USA
1-800-832-4042
ttup@ttu.edu
www.ttup.ttu.edu

This book is dedicated to the first author's
grandparents Edith Theresa and William
Edward Peterson, editors, printers, and
publishers of their day; to Chester M. Rowell, for
his many botanical contributions in the region;
and to his student Jim Blassingame, who
specialized in the study of ferns.

Contents

Illustrations	ix
Preface	xi
Acknowledgments	xiii
Introduction	xv
Ferns and Fern Allies	1
Key to the Trans-Pecos Families of Ferns and Fern Allies	15
Quick Identification of Certain Trans-Pecos Ferns	17
List of Taxa	18
Taxonomic Treatments	22
Selected Glossary	101
Literature Cited	108
Index	111

Illustrations

Fig. 1. Trans-Pecos Texas and adjacent counties, with major geographic features shown. xvii

Fig. 2. The parts of a mature fern. Examples of circinate (fiddlehead) and non-circinate (shepherd's crook) croziers. Examples of sori with true indusia and false indusia. 7

Fig. 3. Frond dissections. 8

Fig. 4. Fern life cycle. Sporophyte generation; mature plant, pinnule, sorus, sporangium, sporangium releasing spores. Gametophyte generation; prothallus, rhizoids, antheridia, archegonia, first root, and first leaf. 11

Fig. 5. *Selaginella* leaves and habits: rosette; erect; prostrate. 23

Fig. 6. *Equisetum laevigatum*, habit; cone. *E. hyemale*, cone. Spores with elaters. 32

Fig. 7. *Ophioglossum polyphyllum*, habit. *O. petiolatum*, habit. Both with enlarged view of vein areoles. 35

Fig. 8. *Anemia mexicana*, habit. 37

Fig. 9. *Adiantum capillus-veneris*, habit; pinnule. 40

Fig. 10. *Argyrochosma microphylla*, habit; rhizome scale; pinnule. *A. limitanea* ssp. *mexicana*, frond; pinnule. 42

Fig. 11. *Astrolepis sinuata*, habit; pinnule; lower scale; upper scale. *A. integerrima*, pinnule; upper scale. *A. windhamii*, pinnule; upper scale. *A. cochisensis*, habit; pinnule; upper scales; lower scale. 46

Fig. 12. *Bommeria hispida*, habit; pinnule. 48

Fig. 13. *Cheilanthes alabamensis*, habit; pinnule. *C. aemula*, frond; pinnule. 53

Fig. 14. *Cheilanthes bonariensis*, habit; pinnule. 54

Fig. 15. *Cheilanthes feei*, frond; pinnule, lower surface, upper surface. *C. villosa*, pinnule, lower surface, upper surface; costal scale. *C. eatonii*, habit; pinnule; costal scale. *C. fendleri*, habit; pinnule. 55

Fig. 16. *Cheilanthes tomentosa*, frond; pinnule; costal scale. *C. horridula*, frond; pinnule, upper surface, lower surface. 57

Fig. 17. *Cheilanthes wrightii*, habit; pinnule. *C. kaulfussii*, pinna; pinnule. *C. lendigera*, frond; pinnule. 58

Fig. 18. *Cheilanthes yavapensis*, pinnule; costal scale.
  C. wootonii, pinnule; costal scale. *C. lindheimeri*, habit;
  pinnule, upper surface, lower surface; costal scale.   59
Fig. 19. *Notholaena grayi*, habit; pinnule; rhizome scale. *N. aliena*,
  pinnule, upper surface, lower surface. *N. aschenborniana*, habit;
  pinnule; rhizome scale.   64
Fig. 20. *Notholaena copelandii*, habit; pinnule. *N. standleyi*, habit;
  pinnule; rhizome scale.   65
Fig. 21. *Notholaena neglecta*, habit; pinnule; rhizome scale.
  N. greggii, frond; pinnule; rhizome scale. *N. nealleyi*, habit;
  pinnule; rhizome scale.   67
Fig. 22. *Pellaea atropurpurea*, frond; rhizome scale. *P. ovata*, frond;
  rhizome scale.   70
Fig. 23. *Pellaea intermedia*, habit; pinnule. *P. cordifolia*, frond;
  stipe with scales; rhizome scale.   72
Fig. 24. *Pellaea wrightiana*, habit; pinnule. *P. truncata*, habit;
  pinnule; rhizome scale. *P. ternifolia*, blade.   73
Fig. 25. *Dennstaedtia globulifera*, pinna; pinnule; sorus.   76
Fig. 26. *Pteridium aquilinum* var. *pubescens*, habit; pinna; pinnules. 77
Fig. 27. *Thelypteris ovata* var. *lindheimeri*, habit; pinnules.   79
Fig. 28. *Asplenium palmeri*, fronds; rooting frond tips. *A. resiliens*,
  frond; fertile pinna. *A. trichomanes*, habit; fertile pinna.   82
Fig. 29. *Asplenium septentrionale*, habit.   83
Fig. 30. *Cystopteris bulbifera*, frond; sorus; fertile pinna with bulblet. 86
Fig. 31. *Dryopteris cinnamomea*, habit; pinnules.   88
Fig. 32. *Dryopteris filix-mas*, habit; sorus; pinna.   89
Fig. 33. *Phanerophlebia umbonata*, habit; pinna; sori. *P. auriculata*,
  pinna.   90
Fig. 34. *Woodsia phillipsii*, pinna. *W. plummerae*, sorus.
  W. neomexicana, habit; sorus; pinna.   93
Fig. 35. *Pleopeltis polylepis* var. *erythrolepis*, habit. *P. riograndensis*,
  habit; pinnule; scale.   95
Fig. 36. *Marsilea vestita*, habit; sporocarp. *M. macropoda*,
  sporocarp. *M. mollis*, sporocarp.   97
Fig. 37. *Azolla mexicana*, habit; leaf.   100
Fig. 38. Glossary illustrations of arrangement, shapes, and tips.   102
Fig. 39. Glossary illustrations of bases, margins, and surfaces.   103

Preface

Tell anyone who knows about ferns that you have been studying these plants in the field, and your listener will assume that you have been tramping around in some humid, heavily vegetated place where the use of mosquito repellent might be a good idea. This is because pteridophytes have the worldwide reputation of being terrestrial plants that are associated with moist habitats, the closer to a rain forest the better.

Studying ferns and fern allies in Trans-Pecos Texas, an arid part of the northern Chihuahuan Desert region, seems somewhat analogous to studying aquatic plants of the region. It should not be a big job, and one should be able to accomplish it by visiting the few perennially moist sites. But wait a moment. There are many interesting arid-adapted pteridophyte species that are found in the desert mountains and on dry slopes of the mesic mountains of the region. And there are many mesic microhabitats even in desertic mountains.

There are north and east slopes, deep canyons, large boulders, shaded woodlands, and ephemeral springs, seeps, and streams where mesic-adapted pteridophytes are found in the Trans-Pecos. Often these are hidden mesic areas, such as the one in upper Limpia Canyon where Jackie Poole discovered the only known United States population of *Pleopeltis polylepis* var. *erythrolepis* (red scale polypody). Many other Trans-Pecos pteridophytes are of restricted distribution in the region, and there are probably more rare treasures to be found in certain microhabitats.

The idea of compiling a book of Trans-Pecos ferns and fern allies is the product of taxonomic work concerning these plants that was being organized by the first author in connection with a larger project, a flora of Trans-Pecos Texas. Her exposure to these plants caused a personal affinity to resurface. She was convinced that an effort should be made to create wider appreciation of the native Trans-Pecos ferns and felt that a book would help accomplish this. Also at stake was the opportunity to promote arid-adapted pteridophytes as plants that have ornamental functions in xeriscaping.

The current work was facilitated by previous treatments of pteridophytes in the *Flora of Texas* (Correll, 1955) and the *Manual of the Vascular Plants of Texas* (Correll and Johnston, 1970). Our work was enhanced by the more recent publication of the *Flora of North America*, volume 2, *Pteridophytes and Gymnosperms* (1993), in which many experts contributed treatments.

The families are arranged in the present work according to the sequence presented in the *Flora of North America*. Within the families, the genera, species, subspecies, and varieties are arranged alphabetically.

The herbarium at Sul Ross State University (SRSC) was our main source of study materials. A few additional specimens were selectively borrowed from other herbaria. In general, botanists do most of their collecting in a few available study locations, such as national and state parks and wildlife management areas, and along established roads, leaving many other areas without adequate representation in the herbarium. This collection pattern can be seen in the lack of herbarium material from the northeastern counties of the Trans-Pecos.

The directions are abbreviated as N, S, E, and W, with C used to indicate "central." Postal Service abbreviations are used for listing the states where a species occurs. Texas (TX) does not appear in these listings when a plant occurs only in the Trans-Pecos area of Texas. Occurrence (or reported occurrence) of a species in Big Bend National Park (B) or Guadalupe Mountains National Park (G) is indicated by the appropriate letter at the beginning of the species description. Elevations are approximate and apply to Trans-Pecos distributions only. Elevations not typical for the species appear in parentheses.

Metric system measurements commonly used in scientific work are employed in keys and descriptions. English measurements have been used for elevations and distances because most southwesterners are accustomed to seeing feet and miles on maps and road signs.

Individual distribution maps are provided for the species. Shading on the maps indicates the areas where each taxon has been documented or may be expected to occur within its preferred habitat. County occurrences that have been reported but not confirmed are indicated by a star.

This publication has been prepared for a wide readership, including amateur naturalists, students, professional botanists, and those just beginning to be interested in these fascinating plants. A selected glossary is included to assist nonspecialists with terms used, and some descriptive terms are illustrated there. Detailed information, such as appears in the section about fern reproduction and life cycles, may be of greater interest to some readers than others, but we hope all readers will use the book in whatever way enhances appreciation of the natural world and the ferns that grow there.

Acknowledgments

We wish to thank the following people who gave invaluable help and assistance in the preparation of this work: for help with fern locations and information, Jim Blassingame, Jean Hardy, Linda Hedges, Denise Louie, Scott Lerich, Barney Lipscomb, and Richard D. Worthington; Billie L. Turner for his unfailing help and encouragement; herbarium research assistants Stephanie Bartel and Patrick Griffith for their support; Petei Zelazny for fern locating and collecting and her wonderful photography; Dorothy Angrist and Jennifer White for their help in preparing the manuscript; Chris Ruggia for assisting with maps and illustrations and sharing his knowledge of computer graphics programs; Bernie Zelazny for sharing his computer and internet expertise; Shirley Powell and Keith Yarborough for their constant support and encouragement; Ellen Carey Bergen-Ruggia and Jane W. Roller for providing the excellent black and white line drawings; and Petei Zelazny for the glossary illustrations and cover photo.

Illustrators

Ellen C. Bergen-Ruggia completed a bachelor of fine arts degree at the University of Texas at Austin in 1991 and conducted graduate studies at Sul Ross State University, also working as a herbarium research assistant. She is a member of the Guild of Natural Science Illustrators, an artist, illustrator, and graphic designer, and co-owner of Vast Graphics, Alpine, Texas.

Jane W. Roller earned a bachelor of science degree at George Washington University in 1938 and a master's degree at Ohio State University in 1943. She has worked as a systematic botanist, plant taxonomist, ecologist, professor, scientific illustrator, and author. A number of her previously published illustrations (Correll, 1955) have been used in the present work.

Petei Zelazny studied at the Dallas Art Institute from 1968 to 1970 and is an artist, illustrator, designer, craftsman, photographer, naturalist specializing in the Big Bend area, and co-owner of Unlimited Limited, Alpine, Texas.

Introduction

Pteridophytes—ferns and their allies—are an inconspicuous part of the landscape in the arid Trans-Pecos region of Texas. Dominant vegetation in this mountain and desert area includes small oak, juniper, and pine trees, medium to small grasses, and many types of desert shrubs. Ferns are most often thought of as growing in moist climates and habitats such as occur in the southeastern half of Texas, but surprising numbers of pteridophyte species occur in the Trans-Pecos. The known Trans-Pecos pteridophyte flora consists of 78 species in 22 genera and 12 families. The largest family is Pteridaceae, with seven genera. The largest genus is *Cheilanthes* with 15 species. Other large genera are *Selaginella* (11 species), *Notholaena* (8 species), and *Pellaea* (7 species).

Mostly these are hardy xeric-adapted species that use water-conserving strategies to survive in dry habitats, as do similarly adapted trees, shrubs, and grasses. Other Trans-Pecos ferns are restricted to relatively moist and shaded microhabitats in mountain canyons, near ponds and ephemeral streams, on north- and east-facing mountain slopes, in water seepage areas, and among large rocks. The bulk of the Trans-Pecos fern flora is biologically related to the xeric and semi-xeric fern flora of central and northwest Mexico (Tryon and Tryon, 1982).

Although ferns and their allies occupy mostly terrestrial habitats, a few are either aquatic (e.g., *Azolla*) or amphibious (e.g., *Marsilea, Equisetum*). In the Trans-Pecos the floating fern, *Azolla*, usually is found on quiet ephemeral streams and natural ponds, while *Marsilea* and *Equisetum* are found at the margins of similar habitats and also at the periphery of some stock tanks.

Trans-Pecos Texas. The unique shape of Texas is easy to recognize on a map of North America. Geographic regions within Texas are not so easy to identify, and for the most part, intrastate regions do not receive standardized designations in popular media. References to "West Texas," for example, may imply the whole region west of Fort Worth and would thus include the geographic center of Texas (in McCullough County), the Panhandle, and far West Texas, the area jutting westward south of the Panhandle. The far western portion of the state is also known as the Trans-Pecos region (fig. 1). It consists of that part of Texas lying west of the well-known Pecos River (*trans* = on the other side), an area approximately the size of the state of Maine.

Topography. The Trans-Pecos is characterized by a mountain and basin topography (Schmidly, 1977; Powell, 1994, 1998), with desert and grassland "seas" surrounding mountain "islands" (Gehlbach, 1981). The Trans-Pecos is part of the Chihuahuan Desert, the largest desert region in North America, which extends from Mexico into southern New Mexico and a small part of southeastern Arizona. About two-thirds of the Chihuahuan Desert lies in Mexico.

The major mountain systems in the Trans-Pecos occur along a northwest-to-southeast axis in the central portion of the region. The highest peaks are found in the Guadalupe Mountains, which extend from south-central New Mexico into northwest Culberson County near the Hudspeth County line. Several peaks in the Texas part of the Guadalupe Mountains exceed 8,000 feet, and the highest elevation in Texas is 8,749 feet at Guadalupe Peak. The Davis Mountains are the largest mountain system in the Trans-Pecos. Most of this system and its highest peak, Mount Livermore (8,382 feet), are in Jeff Davis County. The third highest elevation in the Trans-Pecos, outside the Guadalupe Mountains, is at Emory Peak (7,835 feet) in the Chisos Mountains of southern Brewster County. In western Presidio County, the Chinati Mountains include Chinati Peak at 7,730 feet.

These four mountain ranges and the Glass Mountains all support woodland vegetation from middle to upper elevations, giving way to grassland or desertscrub at lower elevations. Many smaller arid mountains occur throughout the Trans-Pecos west to the Franklin Mountains in El Paso County. These are dominated vegetatively by desertscrub and grassland, with some woodland in more protected habitats. The Trans-Pecos mountains are composed of igneous or sedimentary (mostly limestone) substrates or both. The lowest elevation in the Trans-Pecos is about 1,000 feet at the mouth of the Pecos River in Val Verde County.

Soils. Soils in the Trans-Pecos are exceedingly variable. Rocky substrates predominate in the vicinity of mountains, except where alluvial deposition has filled mountain basins with gravel, sand, and clay. Alluvial deposition is also common in desert habitats where substrate mosaics are characteristic. Desert basins in some areas support highly alkaline soils. Gypsum deposits are evident in the vicinity of ephemeral salt lakes and are exposed elsewhere in the Trans-Pecos. Deep sand deposits are found at certain sites, most notably in El Paso County and just east of the Pecos River in Ward, Crane, Winkler, and adjacent counties.

Climate. Average annual precipitation in the Trans-Pecos is about 12 inches, but ranges from about 8 inches at El Paso up to 20 inches or more at higher elevations in the Davis Mountains. Most rainfall comes in

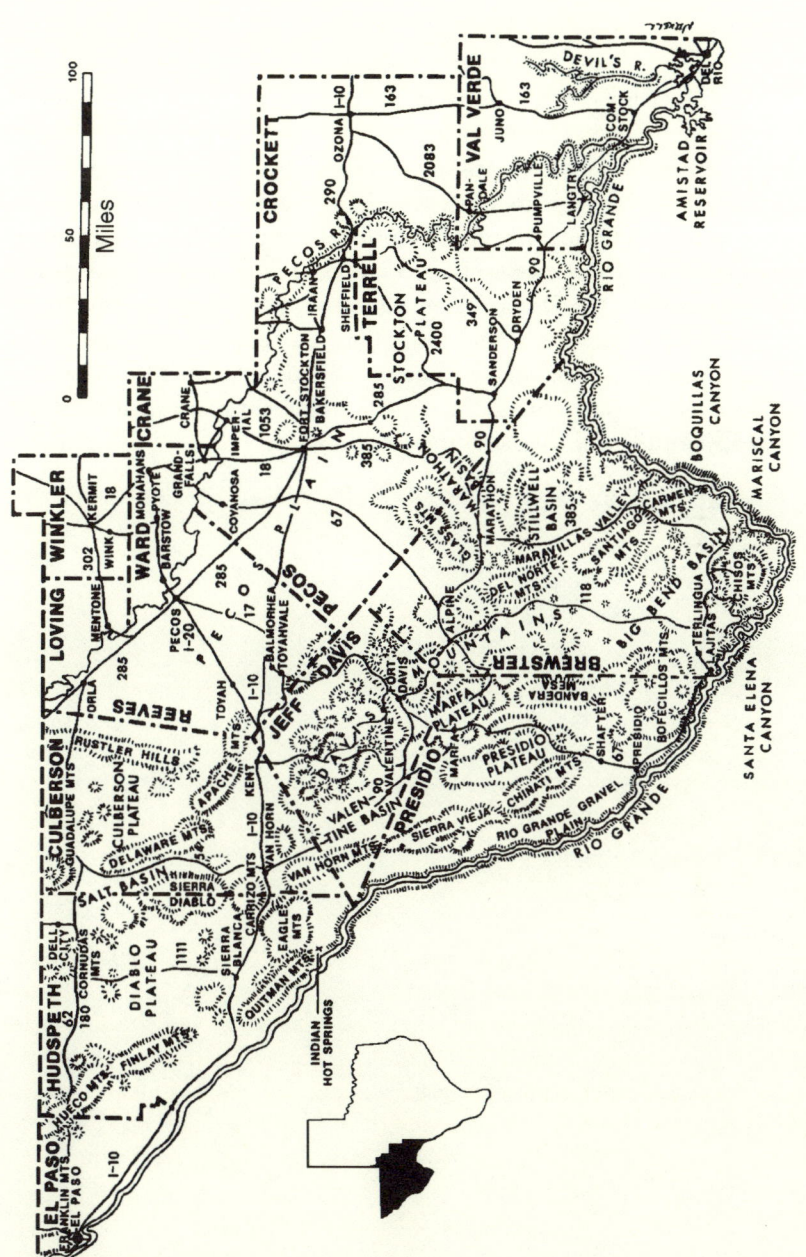

Fig. 1. Trans-Pecos Texas and adjacent counties, with major geographic features shown.

summer thundershowers. Winters and springs are relatively dry. Average temperatures vary widely with elevation. Summer highs range from about 85 to 95°F in the mountains to near 110°F (occasionally near 120°F) in the low desert. Low temperatures in midwinter average about 27–32°F in the mountains, with much lower temperatures, 0–10°F or lower, associated with periodic cold spells.

Vegetation. Woodland, grassland, and desertscrub vegetation in the Trans-Pecos are closely associated with topography, as already mentioned. Characteristic plant communities in the region also are associated with different soil types, such as those that occur with clay, sand, gypsum, and alkaline substrates (Henrickson and Johnston, 1986). Most of the Trans-Pecos species of ferns and fern allies are found in mountain habitats, both the more mesic wooded mountains and the arid desert mountains. Included for purposes of the present treatment are fern species that occur in adjacent counties east of the Pecos River.

Ferns

and Fern Allies of the Trans-Pecos and
Adjacent Areas

Ferns and Fern Allies

The ferns and fern allies are seedless vascular plants that have an evolutionary history dating back over 300 million years. The modern seed-producing plants are the gymnosperms (mainly conifers, such as pines) and angiosperms (flowering plants). Vascular plants are those that have xylem and phloem, the specialized tissues that conduct water and soluble organics through roots, stems, and leaves (when present). In North America north of Mexico there are about 25 families of ferns (often referred to as "true ferns") and five families of the so-called fern allies.

The true ferns, about 11,000 living species, are classified in the phylum (= division) Pterophyta. The fern allies are classified in three phyla: Psilophyta (psilophytes), with two living genera (*Psilotum*, the whisk ferns, and *Tmesipteris*) and several living species; Lycophyta (lycophytes), with 10–15 living genera, including *Lycopodium* (clubmosses), *Selaginella* (spikemosses), and *Isoetes* (quillworts) and about 1,000 living species; and Sphenophyta (horsetails), with a single living genus, *Equisetum* (horsetails or scouring rushes), and 15 living species. Fern allies included in the *Flora of North America* (1993) are conveniently organized by their respective families: Psilotaceae (*Psilotum*), Lycopodiaceae (*Lycopodium*), Selaginellaceae (*Selaginella*), Isoetaceae (*Isoetes*), and Equisetaceae (*Equisetum*). Species of all these families occur in Texas, although only *Selaginella* and *Equisetum* are known to occur in the Trans-Pecos.

Fern Ancestry. The fossil record shows that seedless vascular plants originated during the Devonian period of the Paleozoic era, about 408 to 386 million years ago (Kenrick and Crane, 1997). During the Carboniferous period that followed (about 360 to 286 million years ago), land surfaces were dominated by shallow seas or swamps, and the tropical to subtropical climate was favorable for the year-round growth of plants (Raven et al., 1999). The dominant plants of the Carboniferous were three groups of seedless vascular plants and two groups of seed plants. The seedless plants included lycophyte trees 30 to 110 feet tall; sphenophyte trees or giant horsetails (*Calamites*) reaching 50 feet or more in height; and ferns, including tree ferns up to 25 feet tall or more. The seed plants of the Carboniferous were the so-called seed ferns, now known to be gymnosperms with fernlike leaves, and the cordaites, shrubs and trees up to 45–95 feet tall.

These plants formed great forests that contributed to the "Age of Coal"—the era when coal deposits were initiated during the Pennsylvanian or Upper Carboniferous period (325–286 million years ago). Lycophyte trees apparently dominated the coal-forming swamplands, although the zenith in fern development on the earth, the "Age of the Ferns," also occurred during the late Carboniferous period. Also present in the Carboniferous swamps were herbaceous relatives of the modern seedless vascular plants. By the late Paleozoic, perhaps before 290 million years ago, increasing tropical drought had contributed to the extinction of the lycophyte and sphenophyte trees. Today the herbaceous *Lycopodium, Selaginella,* and *Isoetes* are the only surviving relatives of the Carboniferous lycophytes. *Equisetum* is the only surviving member of the sphenophytes (horsetails), and several groups of modern ferns appear to have descended from those present during the Carboniferous period.

Economic Importance of Ferns

Over the geologic time scale, ferns and fern allies have contributed greatly to the industrial age through the formation of fossil fuels. In the modern world these plants are of relatively minor economic significance, compared to the predominant seed plants (angiosperms and gymnosperms) that are of supreme importance to civilization. Current and historical uses of ferns and fern allies include decoration, medicine, food, and beverage.

Ornamental and Decorative Uses. Ferns are most used today as outdoor and indoor decoration. This is because many species have attractive dissected leaves and because some species thrive in shaded conditions. Serious gardening with ferns began in Britain in the middle of the nineteenth century (Mickel, 1994). Plants were collected from the wild and cultivated in outdoor gardens and indoors in terraria and other containers. As interest in ferns expanded, people gathered on special occasions to visit the countryside in search of interesting species and variant plants. Books appeared concerning the British fern flora, and nurseries specializing in ferns developed large selections of species for sale. The British obsession with ferns was described as "the Victorian fern craze" (Allen, 1969). Not only were ferns being grown, but they were being used as designs on many different household objects, including pottery, silverware, chamber pots, and furniture. A British Pteridological Society was organized in 1891 (pteridology being the study of ferns and fern allies). The British interest in fern cultivation has persisted to the present. Australia and New Zealand also currently have a number of avid fern collectors and growers.

Although the American Fern Society was established in 1893, fern cultivation has not been as popular in the United States as it is in Britain. In America there was considerable early interest in collecting fern specimens and studying the native flora. Fern cultivation has tended toward the tropical species that can be grown indoors anywhere in the country but outdoors only in the warmest climates. More recently, appreciation for cultivating hardy temperate native fern species and hardy temperate species from around the world has increased in the United States. Many regional fern organizations have been founded. In 1954 a fern spore exchange program was initiated through the American Fern Society. Because ferns can be grown from spores, hundreds of different species ultimately were made available to enthusiasts. This approach not only increased the selection of different ferns available to gardeners; it also provided an alternative to the depletion of natural populations through fern gathering. Many books on North American fern identification and cultivation are now available (e.g., Mickel, 1994).

The widespread use of cut fronds (fern leaves) as attractive greenery in floral arrangements is well-known in the United States. Fronds are also used in dried arrangements, featuring either the green color or a sprayed artificial color, often gold or silver. Freeze-dried and painted fronds are popular in some countries. Two spikemoss species, Selaginella lepidophylla and S. pilifera, are collected dry and sold as novelty items. When dry, these small rosette-forming fern allies curl into tight tan or brownish ball-like objects. In this condition they look like dead plants. But when hydrated, the plants rapidly turn green as the stems uncurl to form the spreading rosette habit. Both these Selaginella species are native to the Trans-Pecos, where they are commonly known as resurrection plants.

Trans-Pecos ferns and fern allies have not been much used as ornamentals in the region. Most of the arid-adapted Trans-Pecos ferns have relatively short fronds, but they are nonetheless dissected and attractive when green. As ornamentals these plants hold most promise in xeriscaping, a water-conserving form of landscaping that has gained acceptance in the desertic southwestern United States. Examples of the use of Trans-Pecos native ferns and Selaginella lepidophylla can be seen in a formal cactus garden on the Sul Ross State University campus in Alpine.

Medicinal Uses. Historically, ferns have been used in folk medicine in many parts of the world. Usually infusions or decoctions of rhizomes or fronds were taken internally, but solutions extracted from these organs, or in some cases all parts of the plants, including spores, were used topically as a paste or poultice. Some of the folk remedies are considered efficacious by today's standards, but modern use of ferns in medicine is of minor and local significance. Preparations from certain species of ferns have been used to expel intestinal worms, to treat dysentery, to relieve pain from bites and stings, to soothe burns and bruises, and to cure asthma and rheumatism (Jones, 1987). Many ferns and fern allies, including *Selaginella lepidophylla*, have been shown to have diuretic properties. Preparations from other species have been used to treat ulcers and to control bleeding. Where these are known, medicinal reputations associated with Trans-Pecos plants are cited in the text.

Food Uses. Young leaves and rhizomes of many fern species are eaten, either raw or cooked, in many parts of the world, including the United States. Edible fronds are those that are in the young, tender stage of unrolling, a stage known as fiddleheads or croziers. Usually the fiddleheads are cooked by boiling them in salty water for one hour or less or by steaming them until the tissue is soft. Fiddleheads are cooked whole or sliced. Raw fiddleheads make a crisp, but mucilaginous, addition to salads. In the United States, *Matteuccia struthiopteris* (ostrich fern, the state vegetable of Vermont) and *Osmunda cinnamomea* (cinnamon fern) produce fiddleheads that are considered to be delicacies. The canning of ostrich fern fiddleheads has been commercialized in New England and adjacent Canada (*Flora of North America*, 1993).

The starchy rhizomes of numerous fern species are eaten in other parts of the world. In most cases, the rhizomes are roasted whole before they are eaten, but those of some species are soaked and pounded before being cooked. Pith in the trunks of certain tree ferns is cooked and eaten in New Guinea, New Zealand, New Caledonia, the Philippines, Australia, and Hawaii. The palatability of the fiddleheads and rhizomes of Trans-Pecos ferns is not known.

Poisonous Ferns. A few species of ferns and fern allies are known to be toxic to livestock. For example, the Australian mulga fern, *Cheilanthes sieberi*, is reported to be poisonous to sheep and cattle. *Equisetum arvense* (horsetail) may be toxic to horses. Bracken ferns (*Pteridium aquilinum*) are not only poisonous to livestock but also produce carcinogenic substances that reportedly cause cancer in humans when fiddleheads are ingested in large quantities.

In the Trans-Pecos, *Astrolepis cochisensis* (jimmyfern) is known to be toxic to sheep and goats. A related taxon in California also is reported to be poisonous to sheep, as is the Trans-Pecos species *Argyrochosma microphylla*.

Other Uses. A tea is brewed from the fresh or dried leaves of *Dryopteris fragrans* (fragrant wood fern), *Pellaea mucronata*, and *Adiantum capillus-veneris* (maidenhair fern), the last of which occurs in the Trans-Pecos. Tea from the leaves of the maidenhair fern also is reported to have medicinal properties.

Tree fern trunks have been used in the construction of buildings and fences. The scales or hairs of certain tree ferns are used as pillow stuffings. The matty, tough root systems of various species of *Osmunda* and certain other fern genera are used as part of a substrate for growing epiphytic orchids. The flexible stems and leaf rachises of some climbing ferns are used to make ropes, baskets, chairs, partitions in buildings, and fish traps.

Ferns in their natural habitats serve the important function of preventing erosion. Certain weedy ferns have been employed in the stabilization and reclamation of dirt banks, road cuts, and deep sand. Rice farmers in Asia reportedly encourage, or at least tolerate, the growth of the floating fern *Azolla* in their paddy fields because the plants are known to contain the nitrogen-fixing symbiont *Anabaena azollae* (a cyanobacterium), which helps fertilize the soil. Various additional miscellaneous uses of ferns of the world are discussed by Jones (1987).

Fern Morphology

As vascular plants, mature ferns consist of roots, stems, and leaves. The leaves, commonly known as fronds, are the dominant organ. In lay terms, the typical fern plant might be described as a tuft of leaves emerging from the ground. The stem is relatively inconspicuous, either just beneath or on the ground surface, generally horizontal and creeping. Stems give rise to leaves that carry out photosynthesis and to adventitious roots that absorb water and nutrients and help anchor the plant.

Stems. The horizontal or ascending fern stems are called rhizomes. New leaves are initiated from the growing tip of the rhizome. When horizontal, the rhizome usually is short-creeping, resulting in mature plants with closely clumped fronds. In plants with long-creeping rhizomes, the leaves are more distant, resulting in a diffuse clump of fronds. Ascending rhizomes may form a caudex (see glossary), or be cormlike, as in *Ophioglossum*, and often produce densely clumped fronds.

Roots in Trans-Pecos ferns usually are slender and wiry and arise along the length of the rhizome. In *Ophioglossum* the roots are slender but somewhat fleshy.

Fern rhizomes and fronds may be covered with distinctive hairs or scales. Hairs are slender structures that are only one cell wide. In some ferns, the hairs are multicellular (septate), while in others, such as *Thelypteris*, the hairs are one-celled (simple). Some ferns have glandular hairs that secrete volatile oils, as in *Dennstaedtia*, or that produce white or yellow wax (farina), as in *Notholaena*. Scales are many cells wide and one cell thick (monostromatic). Some scales may be slender and resemble hairs. Other scales may be more or less lance-shaped. Especially distinctive clathrate scales are produced by the genus *Asplenium* (the spleenworts). Clathrate scales often exhibit translucent colors and thick-walled cells that are evident under a hand lens and have been likened to stained glass.

Fronds. Fern leaves are megaphylls. Morphologically, a megaphyll is a kind of leaf that produces a "leaf gap" in the vascular tissue; that is, a break in vascular tissue at the point where it diverges from the stem and enters the leaf. Megaphylls are the kind of leaves also produced by the modern seed plants, the gymnosperms and angiosperms. Megaphyll means "large leaf," and indeed the leaves of megaphyllous plants typically are larger than the usually scalelike leaves of the microphyllous fern allies. But the morphological description of a microphyll is a leaf that does not leave a leaf gap; it has a single unbranched vein that does not leave a gap in the vascular pattern of the stem.

Young fern leaves are distinctive. They are coiled in bud (fig. 2), which is known as exhibiting circinate vernation. In their rolled-up condition, young fern leaves are called fiddleheads or croziers, as earlier noted. As the leaves mature, from the base to the tip, they gradually unroll from the fiddlehead, exhibiting leaf structure typical of the mature frond. Some ferns produce non-circinate young leaves that are merely hooked at the apex, like a shepherd's crook (fig. 2). A few angiosperms also produce circinate leaves.

A mature frond exhibits a stipe and a blade (fig. 2). The stipe is the slender petiole or stalk by which the blade is attached to the stem. The blade is the expanded portion of the frond and is either simple or dissected. The pattern of vascular bundles in the stipe, easily observed in fresh cross section, is useful in fern taxonomy but not much alluded to in the present treatment.

Frond morphology is highly significant in fern taxonomy. In relatively few ferns the blade may be simple or entire rather than divided or lobed (fig. 3). In most ferns there is some degree of dissection or division of the blade, and there are specific terms for each degree of dissection. A pinnatifid blade is deeply lobed, with the divisions extending from about one-quarter or one-half to nearly all the way to the rachis (fig. 3). The

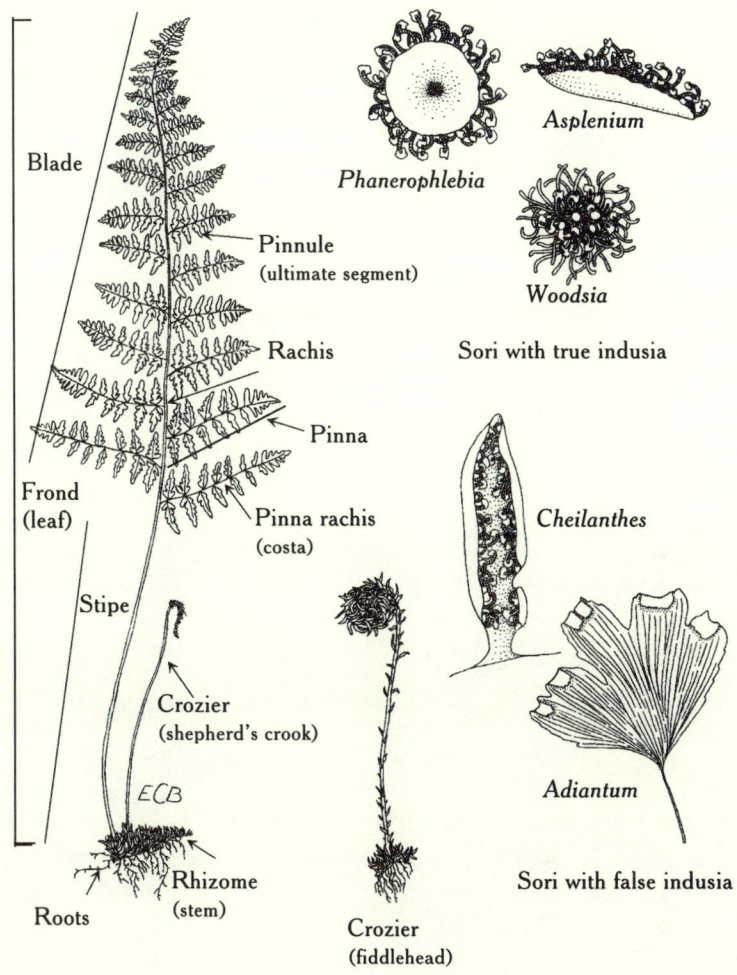

Blade

Pinnule
(ultimate segment)

Rachis

Pinna

Frond
(leaf)

Pinna rachis
(costa)

Stipe

Crozier
(shepherd's crook)

ECB

Rhizome
(stem)

Roots

Crozier
(fiddlehead)

Phanerophlebia

Asplenium

Woodsia

Sori with true indusia

Cheilanthes

Adiantum

Sori with false indusia

Fig. 2. The parts of a mature fern. Examples of circinate (fiddlehead) and non-circinate (shepherd's crook) croziers. Examples of sori with true indusia and false indusia.

rachis is the axis that bears the leaflets, essentially an extension of the stipe. A *pinnate* blade has leaflets, not lobes, with each leaflet attached to the rachis by a slender stalk.

Blades more divided than once-pinnate are termed *bipinnate* or *tripinnate* (fig. 3). If leaflets (pinnae) are lobed, they are also pinnatifid (e.g., pinnate-pinnatifid or bipinnate-pinnatifid in fig. 3). If leaflets are further divided into pinnules, then a blade is described as bi- or tripinnate,

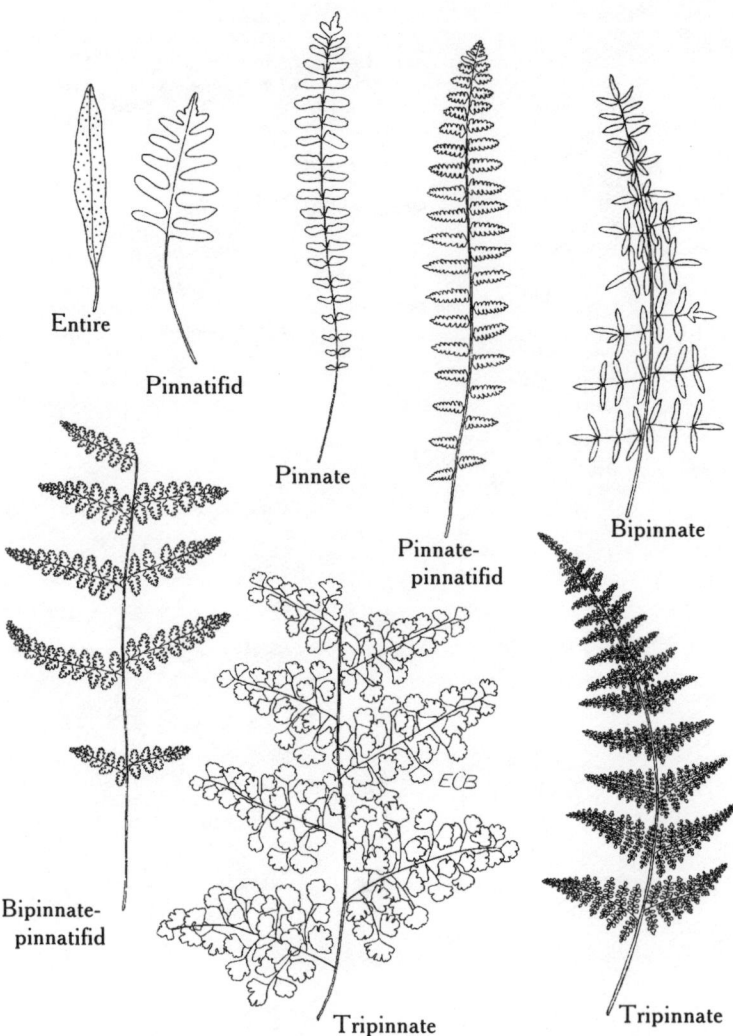

Fig. 3. Frond dissections.

depending upon whether the blade is twice or three times dissected. In some ferns, leaves are four or five times pinnate.

  Frond venation, seen on the surface of the frond, is potentially important in fern taxonomy. Ordinarily there is a midvein of leaflets and smaller veins extending toward the margins. Free veins are found in most North American ferns (Mickel, 1979)—that is, the veins are separate, not forming a network. Networks of united veins are distinctive in some ferns.

Leaf surfaces of ferns may have scales, hairs, or waxy particles on the upper (adaxial) or the underneath (abaxial) surfaces or both. The hairs may be simple, branched, or stellate. Scales may be distinctive in appearance. Leaves of many fern species are naked or mostly so (see figs. 38 and 39 in glossary for illustrations of terms describing leaf shapes, arrangements, bases, margins, and surfaces).

Fern Reproduction

Ferns have what is known as an alternation of generations (Bold et al., 1987), in which there are two reproductive stages very different in outward structure and appearance, chromosome number, and function. The sporophyte generation, which has a full complement of chromosomes, is what we generally think of as the actual fern plant with leaves, stems, and roots. Usually the sporophyte has two sets of chromosomes, known as diploid ($2n$). In many ferns, however, there can be additional sets of chromosomes present, known as higher ploidy levels. Examples of other ploidy levels could be three sets of chromosomes, known as triploid ($3n$); four sets, known as tetraploid ($4n$); or even higher levels.

The sporophyte generation, with its full complement of chromosomes ($2n$), normally produces spores that have half the number of chromosomes, known as haploid (n). These spores can then give rise to the haploid gametophyte generation, which continues to have half the number of chromosomes of the sporophyte generation. The structures of the gametophyte generation are seldom seen because they are very small (5–6 mm). These tiny structures are often green and heart-shaped. As the gametophyte generation develops, it produces both the male and female sex organs, which then may promote sexual reproduction and once more create the sporophyte generation with its two sets of chromosomes.

Reproductive Structures and General Life Cycle. Sporangia (cases that hold the spores), when present, are borne on the underside of fern leaves. Sporangia usually are produced in clusters known as sori (singular, sorus). Mature sori look like tiny brownish spheres and are evident to the naked eye. For the inexperienced observer, interpreting mature sori as insect infestation on the underside of fern leaves is not an uncommon first impression. Characters of the sori, particularly their shape and arrangement on the pinnae, are of major significance in fern identification. Usually the sori are associated with veins and located between the midvein and margin of the blade or leaflet (that is, they are medial). In other ferns, sori may be near the margin (marginal) or near the midvein. In many fern species, the development of young sori is associated with the formation of a covering flap of tissue called the indusium (e.g., *Woodsia*, *Asplenium*, *Phanerophlebia* in fig. 2). The indusium sometimes withers or shrinks with sorus maturation. In certain ferns, marginal sori are covered by the re-curved leaf margin, which is known as a false indusium (*Adiantum*, *Cheilanthes* in fig. 2). Indusium and false indusium morphology, when present, are important in fern identification.

Although individual fern sporangia are visible to the naked eye, micro-scopic aid is necessary to see details of the structure (fig. 4). A sporangium in most ferns is a thin-walled spore case, usually attached to the leaf by a stalk. Arching over the top of each spore case is a single row of specialized thick-walled cells, the annulus, which functions in opening the sporangium and later in spore dissemination (fig. 4). Fern sporangia of different spe-cies produce characteristic numbers of spores. In primitive ferns, 500 to thousands of spores are produced; in other species, there may be as few as 16 to 32 spores per sporangium. The most common number of spores in advanced ferns is 64. Spore shape and spore wall ornamentation are other important taxonomic characters in pteridophytes. Usually members of the same genus or family produce the same spore shapes. Spore wall sculptur-ing may be useful in identifying species or genera.

Fern spores that reach a favorable substrate may germinate to produce the gametophyte, the gamete-producing generation. The gametophyte it-self is known as a prothallus (pl. prothalli, fig. 4). The prothallus is small, usually about 5–6 mm in diameter, dorsiventral, generally heart-shaped, green, and photosynthetic. The central region of the prothallus is several cells thick, while the "wings" are monostromatic (one cell layer thick). Anchoring rhizoids are produced underneath and near the base of the prothallus. Male (antheridia) and female (archegonia) sex organs are produced on the underside of mature and functional prothalli (fig. 4). Antheridia are located in the central region near the base of the prothallus and among the rhizoids. Archegonia are located in the central region near the apical notch. Mature antheridia release multiflagellate sperm, each with one set of chromosomes, that swim to the archegonia. After they swim down the neck of the flask-shaped archegonium, fertilization is effected through the fusion of the sperm nucleus with the egg nucleus, which also contains one set of chromosomes. The fertilized egg, now containing two sets of chromosomes, is known as the zygote and is the first cell of the new sporophyte generation. Although many archegonia are present, usually only one fertilization takes place on each prothallus, or at least only one sporophyte usually develops.

In juvenile sporophyte (sporeling) development, the zygote undergoes many divisions that result in the formation of an embryo. The embryonic

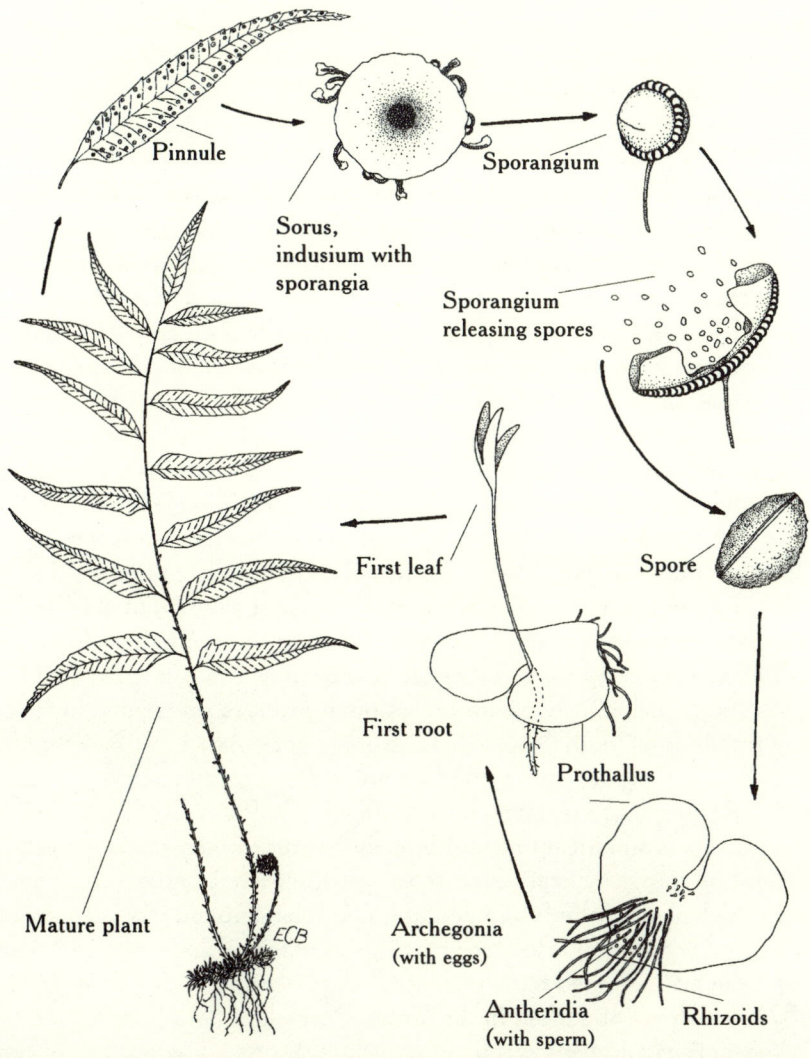

Fig. 4. Fern life cycle. Sporophyte generation; mature plant, pinnule, sorus, sporangium, sporangium releasing spores. Gametophyte generation; prothallus, rhizoids, antheridia, archegonia, first root, and first leaf.

structure consists of a foot, root, stem, and leaf. The foot is a relatively undifferentiated mass of tissue that grows into and remains embedded in the prothallus. It absorbs nutrition that is used in supporting early growth of the sporophyte. The first leaf of the sporophyte—not at all typical in appearance or similar to those of the adult sporophyte—grows up through the apical notch (fig. 4) and functions in photosynthesis. The embryonic stem ultimately elongates, giving rise to roots and leaves that are at first circinate and then develop into a form characteristic of the species. The prothallus withers as the sporophyte becomes self-sufficient.

Heterosporous Fern Life Cycle. The generalized life cycle described is for homosporous true ferns, in which all of the spores produced in sporangia are alike. Ferns of the families Marsiliaceae, Salviniaceae, and Azollaceae produce spores of two sizes and types, microspores and megaspores, and are therefore heterosporous. Germination of the microspores results in the development of a small prothallus that produces only antheridia. After germination, the megaspores produce only archegonia. Fertilization occurs in the manner already described for homosporous ferns, when functional microspores and megaspores are deposited in close proximity. The fern allies *Selaginella* and *Isoetes* also are heterosporous.

Asexual Reproduction. At least two types of specialized asexual reproduction are widespread in ferns and are common in some groups. Asexual or vegetative reproduction in dry habitats is advantageous because it eliminates the need for water to convey flagellated male gametes during sexual reproduction. Apogamy is a process whereby vegetative buds are developed on the midregion or on the margin of the prothallus. Development of a vegetative bud gives rise to a sporophyte in a manner that bypasses sexual fusion. Apogamous ferns often produce spores that have the same ploidy level as the sporophyte. Spores germinate to form a diploid prothallus, and the vegetative budding process regenerates a diploid sporophyte (agamospory; Gastony and Windham, 1989).

Apogamy is significant in fern biology because it allows for the establishment of sterile hybrids, even those with uneven numbers of chromosomes, such as in triploids and pentaploids. Even haploid ferns can proliferate through apogamy. Relatively high percentages of apogamy are known to occur in arid-adapted ferns such as *Cheilanthes* and *Notholaena* (Jones, 1987), which are abundant in the Trans-Pecos fern flora.

Apospory is another vegetative reproductive process in ferns. This occurs when a prothallus grows directly from a leaf, with spores being eliminated from the process. Aposporous prothalli can form normal antheridia and archegonia.

Fern Allies Reproduction

The fern allies are a polymorphic (having many forms) group of seedless vascular plants that reproduce by disseminating spores. It seems obvious that the fern allies, although structurally different in several respects, are related at least distantly to the true ferns. The fern allies include *Psilotum* and *Tmesipteris* (whisk ferns, or fork ferns), *Lycopodium* and related genera (clubmosses), *Selaginella* (spikemosses), *Isoetes* (quillworts), and *Equisetum* (horsetails, or scouring rushes). Clubmosses and spikemosses produce microphyllous leaves. Quillworts have peculiar linear leaves, while those of the horsetails are scalelike. Life cycles of the fern allies are similar to the fern life cycle. Plants produce sporangia, giving rise to spores that germinate to form an independent prothallus or prothalli with sex organs. After fertilization, a new sporophyte is produced. Like true ferns, the fern allies are homosporous, except for the heterosporous *Selaginella* and *Isoetes*. The genera *Selaginella* and *Equisetum* and some species in the genus *Lycopodium* produce sporangia in specialized sporophyll clusters called strobili. In *Isoetes*, sporangia are associated with swollen leaf bases.

Among the fern allies, only *Selaginella* and *Equisetum* are found in the Trans-Pecos. Species of *Psilotum* and *Isoetes* and members of the family Lycopodiaceae are found elsewhere in Texas.

Hybridization

Natural interspecific (betweeen species) hybridization, often associated with polyploidy (having three or more sets of chromosomes), has played a major role in the evolution of ferns. Some of these hybridizations between species are exceedingly complex (Gastony, 1986) and are not discussed here. Many fern hybrids are sterile, producing mainly abortive spores, but sterile hybrids may be perpetuated through asexual reproduction or through the occasional production of viable spores. In some fern groups, speciation has followed interspecific hybridization—usually associated with polyploidy—and subsequent formation of fertile populations. Analysis of hybridizations and chromosome numbers has been important in modern studies of ferns.

Fern Distribution and Habitat

Ferns and their allies are widely distributed in the world, being most common in moist habitats, especially in mountain regions of tropical and subtropical forests. They are common in temperate zones. Ferns occur commonly from sea level to about 11,500 feet (in the Andes, New Guinea, and the Himalayas) and rarely to about 14,000 feet in the Andes. Pteridophytes are uncommon or absent in the driest and coldest areas. About 75 percent of the fern species in the world occur in two great regions (Tryon and Tryon, 1982). One distributional center is in the Americas, from southern Mexico through Central America and into South America, particularly

in the Andes from Venezuela south to Bolivia, and also in the Greater An-
tilles (mostly in Cuba, Jamaica, and Hispanola). The other center is southeastern Asia and Malaysia. Africa has comparatively few pteridophyte
species.

In the Trans-Pecos, most fern species occur in the mesic mountains and
in mesic microhabitats in the more arid mountains. In the eastern Trans-
Pecos, ferns are most common in woodlands and in mesic canyons. Most
mountain ranges support at least several species of arid-adapted ferns, even
on relatively dry slopes.

Arid Adaptations

A majority of Trans-Pecos ferns exhibit water-conserving strategies.
These xeric adaptations include relatively small fronds that are shaded
with hairs, scales, or waxy coatings. The fronds may wither and curl during
dry periods or die back to subterranean rhizomes that themselves are
drought-tolerant. It is apparent that these arid-adapted ferns have special
physiological tolerances to minimal water. In dry periods, their curled
fronds may appear completely desiccated, and they can be broken and
crumbled with ease. When rehydrated under natural or artificial conditions, however, the dried leaves may freshen up and turn green overnight
(poikilohydry). Most species of *Cheilanthes, Notholaena,* and *Astrolepis*
are arid-adapted, as are the better known spikemoss resurrection ferns,
Selaginella lepidophylla and *S. pilifera.* Ferns native to dry regions may
also bypass sexual reproduction and undergo a higher frequency of apogamous vegetative reproduction (Jones, 1987), but the extent of apogamy in
Trans-Pecos ferns is yet to be documented.

Key to the Trans-Pecos Families of Ferns and Fern Allies

1. Leaves needle- or scalelike, with one unbranched vein; sporangia (spore cases) borne in cones or leaf axils . . . (2).
1. Leaves usually broader, not needlelike; veins branching; sporangia on leaves or in nutlike sporocarps (separate thick-walled structures) . . . . (3).

2 (1). Plants rushlike, sporangia borne in terminal cones (at the tip or apex of the stems) made up of polygonal, umbrellalike segments; stems jointed, grooved, mostly hollow . . . . 2. Equisetaceae.
2. Plants mosslike, sporangia borne in axils (angles between leaves and stems) of leaflike sporophylls that make up more or less 4-sided cones (strobili); stems not jointed, grooved, or hollow . . . . 1. Selaginellaceae.

3 (1). Plants aquatic, floating or rooting in mud . . . . (4).
3. Plants usually terrestrial (on land) or on rock . . . . . . (5).

4 (3). Plants floating in water; fronds 2-lobed, less than 5 mm wide . . . . 12. Azollaceae.
4. Plants usually rooting in mud; fronds 4-pinnate, (resembling a 4-leaf clover), more than 5 mm wide . . . . 11. Marsileaceae.

5 (3). Fronds divided into a vegetative portion and one or two erect fertile spikelike portions . . . . (6).
5. Fronds not divided as above, lacking a separate erect fertile portion arising from the stipe (petiole) . . . . (7).

6 (5). Erect fertile part, one per frond; stem tuberous, erect, 25 mm or more beneath the ground surface . . . 3. Ophioglossaceae.
6. Erect fertile parts, two per frond; stem a creeping rhizome (modified stem) at, or near, the ground surface . . . . 4. Anemiaceae.

7 (5). Sori without a true indusium (protective covering) but instead
        partially covered by a false indusium made by the folding over
        of pinnae (leaf) edges, or overlapping scales may be present
        . . . . (8).
    7. Sori with a true indusium. . . . . (9).

8 (7). Blades entire (not divided or lobed) or pinnatifid (divided
        into narrow lobes); false indusium of overlapping, peltate
        scales (shieldlike, attached toward the middle); farina (mealy,
        powdery substance on the pinnae undersurface) absent
        . . . . 10. Polypodiaceae.
    8. Blades once-pinnate or more; false indusium of folded-over
        pinnae edges or lacking; scales absent or basifixed (attached at
        the base); farina present or absent . . . . . . 5. Pteridaceae.

9 (7). Sori marginal (at the leaf edges) in cuplike indusia (very
        large, rare ferns of limestone sinkholes, Val Verde Co.), or
        sori in a marginal band; underside of blades with abundant,
        lax, spreading hairs (rare ferns of the Davis Mts., Jeff Davis
        Co.). 6. Dennstaedtiaceae.
    9. Sori submarginal to medial, not marginal; characters not as
        above . . . . (10).

10 (9). Sori elongate, oblique (slanting at an angle) to midrib along
        veins. 8. Aspleniaceae.
    10. Sori round, not oblique to midrib along veins . . . . . . (11).

11 (10). Blade with numerous, tiny, needle-shaped hairs; indusium
        kidney-shaped, attached at the sinus (depression)
        . . . . 7. Thelypteridaceae.
    11. Blade without numerous, tiny, needle-shaped hairs; indusium
        variously shaped and attached . . . . . . 9. Dryopteridaceae.

Quick Identification of Certain Trans-Pecos Ferns

Habit
Floating in water, leaves tiny . . . . *Azolla.*
Mosslike . . . . *Selaginella.*
Resurrection rosette or appearing like a dried bird's nest
. . . . *Selaginella lepidophylla* or *S. pilifera.*
Rushlike. . . . . *Equisetum.*

Height
Over 2 m tall . . . . *Dennstaedtia globulifera.*

Blades
Grasslike . . . . *Asplenium septentrionale.*
Once-pinnate to pinnate-pinnatifid (fig. 3), undersurface covered with
    ciliate scales (with a marginal fringe of hairs) . . . . *Astrolepis.*
Pentagonal (5-sided), star-shaped, undersurface covered with greenish
    yellow farina. . . . . *Notholaena standleyi.*
Pentagonal, star-shaped, undersurface hairy . . . . *Bommeria hispida.*
Pinnate-pinnatifid, undersurface covered with tomentose (matted,
    woolly), usually golden hairs . . . . *Cheilanthes bonariensis.*
Resembling a 4-leaf clover . . . . *Marsilea.*
With an erect spore-bearing spikelike portion, sterile blades entire
. . . . *Ophioglossum.*
With an erect spore-bearing spikelike portion, sterile blades pinnate
. . . . *Anemia.*
Undersurface covered with farina . . . . *Notholaena* or *Argyrochosma.*

Ultimate segments (leaflets farthest from the pinna rachis)
Tiny, undersurface covered with silver farina
. . . . *Argyrochosma limitanea* ssp. *mexicana.*
Tiny, undersurface without silver farina
. . . . *Argyrochosma microphylla.*

Sori
Elongate, arranged obliquely to the midrib along the veins . *Asplenium.*

Indusium
Kidney-shaped, hairs numerous, tiny, needle-shaped
. . . . *Thelypteris ovata* var. *lindheimeri.*

List of Taxa

1. Selaginellaceae (Spikemoss Family)
1. *SELAGINELLA*
   B   1. *Selaginella arizonica*
   B   2. *Selaginella lepidophylla*
   B,G   3. *Selaginella mutica*
           Selaginella mutica var. *mutica*
           Selaginella mutica var. *limitanea*
   B   4. *Selaginella xneomexicana*
   B,G   5. *Selaginella peruviana*
   B,G   6. *Selaginella pilifera*
         7. *Selaginella rupincola*
   B   8. *Selaginella scopulorum*
   B   9. *Selaginella underwoodii*
   B   10. *Selaginella viridissima*
   B,G 11. *Selaginella wrightii*

2. Equisetaceae (Horsetail, Scouring Rush Family)
1. *EQUISETUM*
         1. *Equisetum xferrissii*
         2. *Equisetum hyemale* ssp. *affine*
   B,G   3. *Equisetum laevigatum*

3. Ophioglossaceae (Adder's Tongue Family)
1. *OPHIOGLOSSUM*
         1. *Ophioglossum petiolatum*
         2. *Ophioglossum polyphyllum*

4. Anemiaceae (Anemia Family)
1. *ANEMIA*
   B   1. *Anemia mexicana*

5. Pteridaceae (Maidenhair Fern Family)
1. *ADIANTUM*
   B,G   1. *Adiantum capillus-veneris*

2. ARGYROCHOSMA

B,G   1. *Argyrochosma limitanea* ssp. *mexicana*
B,G   2. *Argyrochosma microphylla*

3. ASTROLEPIS

B,G   1. *Astrolepis cochisensis*
B,G   2. *Astrolepis integerrima*
B,G   3. *Astrolepis sinuata*
B,G   4. *Astrolepis windhamii*

4. BOMMERIA

B   1. *Bommeria hispida*

5. CHEILANTHES

        1. *Cheilanthes aemula*
B,G   2. *Cheilanthes alabamensis*
B   3. *Cheilanthes bonariensis*
B,G   4. *Cheilanthes eatonii*
B,G   5. *Cheilanthes feei*
        6. *Cheilanthes fendleri*
B   7. *Cheilanthes horridula*
B   8. *Cheilanthes kaulfussii*
B.   9. *Cheilanthes lendigera*
B   10. *Cheilanthes lindheimeri*
B   11. *Cheilanthes tomentosa*
B,G 12. *Cheilanthes villosa*
       13. *Cheilanthes wootonii*
B   14. *Cheilanthes wrightii*
       15. *Cheilanthes yavapensis*

6. NOTHOLAENA

B   1. *Notholaena aliena*
        2. *Notholaena aschenborniana*
B   3. *Notholaena copelandii*
B   4. *Notholaena grayi* ssp. *grayi*
B   5. *Notholaena greggii*
B   6. *Notholaena nealleyi*
        7. *Notholaena neglecta*
B,G   8. *Notholaena standleyi*

7. PELLAEA

B,G   1. *Pellaea atropurpurea*
B   2. *Pellaea cordifolia*
B,G   3. *Pellaea intermedia*
B   4. *Pellaea ovata*
B   5. *Pellaea ternifolia*

           6. *Pellaea truncata*
B,G   7. *Pellaea wrightiana*

6. Dennstaedtiaceae (Dennstaedtia Family)
1. *DENNSTAEDTIA*
           1. *Dennstaedtia globulifera*
2. *PTERIDIUM*
B   1. *Pteridium aquilinum* var. *pubescens*

7. Thelypteridaceae (Marsh Fern Family)
1. *THELYPTERIS*
B   1. *Thelypteris ovata* var. *lindheimeri*

8. Aspleniaceae (Spleenwort Family)
1. *ASPLENIUM*
           1. *Asplenium palmeri*
B,G   2. *Asplenium resiliens*
B   3. *Asplenium septentrionale*
           4. *Asplenium trichomanes* ssp. *trichomanes*

9. Dryopteridaceae (Wood Fern Family)
1. *CYSTOPTERIS*
G   1. *Cystopteris bulbifera*
G   2. *Cystopteris reevesiana*
G   3. *Cystopteris utahensis*
2. *DRYOPTERIS*
           1. *Dryopteris cinnamomea*
           2. *Dryopteris filix-mas*
3. *PHANEROPHLEBIA*
G   1. *Phanerophlebia auriculata*
B   2. *Phanerophlebia umbonata*
4. *WOODSIA*
B,G   1. *Woodsia neomexicana*
B,G   2. *Woodsia phillipsii*
           3. *Woodsia plummerae*

10. Polypodiaceae (Polypody Family)
1. *PLEOPELTIS*
           1. *Pleopeltis polylepis* var. *erythrolepis*

B   2. *Pleopeltis riograndensis*

11. Marsileaceae (Water Clover Family)
1. *MARSILEA*
       1. *Marsilea macropoda*
       2. *Marsilea mollis*
       3. *Marsilea vestita*

12. Azollaceae (Azolla Family)
1. *AZOLLA*
       1. *Azolla mexicana*

Taxonomic Treatments

1. **Selaginellaceae** Willk., Spikemoss Family

The spikemoss family is diverse, with members ranging from the small, creeping, mosslike plants of the Trans-Pecos to larger, erect, and frondlike tropical representatives. In the Trans-Pecos, the small perennial selaginellas usually grow on rock or in barren, rocky soil. The stems are prostrate to erect, rosette- (fig. 5) or mat-forming, slender and leafy, and often have rhizophores (elongate, wiry rootlike structures) at the nodes. The one-nerved leaves are small, numerous, linear, lanceolate, elliptic, or deltate in shape and are often tipped with a hairlike bristle. The sporophylls (fertile leaves) form a compact four-sided cone at branch apexes (tips), with solitary one-celled sporangia occurring in the axils. The sporangia are of two kinds: megasporangia containing one–four large megaspores and microsporangia containing numerous microspores.

In the desert regions, selaginellas are plants of semiarid locales where there is little other vegetation. They survive desiccation (drying) for long periods and revive quickly after receiving water. *Selago* is an ancient name for clubmoss, which the selaginellas resemble. The family contains only one genus with over seven hundred species occurring worldwide, mostly in tropical and subtropical regions. Thirty-eight species occur in the United States, with about thirteen species in Texas and eleven species in the Trans-Pecos.

1. *SELAGINELLA* P. Beauv., SPIKEMOSS

Two species of Trans-Pecos selaginella have the rosette or resurrection fern habit, with the hairy *S. pilifera* generally growing in the mountains and the more common *S. lepidophylla* occurring throughout much of the area. The mat-forming selaginellas are small plants, and different species may look similar to the unassisted eye. Two or even three separate species may be found growing in the same mat, making identification even more of a challenge. Some of the prostrate species (growing flat on the ground; fig. 5) have dorsiventral stems with leaves that are larger on the underside and appear to sweep upward. Species without dorsiventral stems have similar-sized leaves, which are arranged symmetrically around the stem. Many of the identifying characters used in the following key are best seen under at least 10-power magnification.

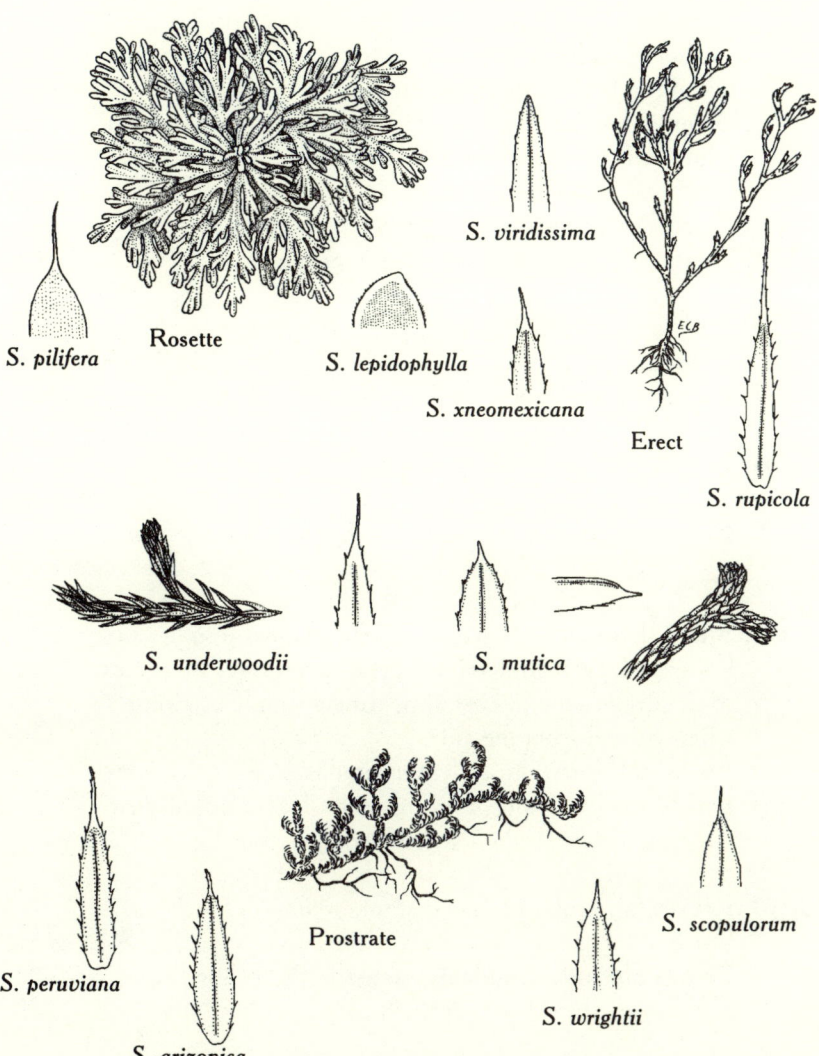

Fig. 5. *Selaginella* leaves and habits: rosette; erect; prostrate.

Key to the Species

1. Plants forming conspicuous rosettes . . . . (2).
1. Plants not rosettelike, usually forming dense mats . . . . (3).

2 (1). Stem leaves ovate-deltoid to deltoid, apex (tip) rounded
        to short cuspidate (tapering gradually to a rigid tip),
        without a bristle, margins often white . . . 2. S. *lepidophylla*.
   2. Stem leaves lanceolate, apex acute with a long,
        white terminal bristle, margins green . . . . . . 6. S. *pilifera*.

3 (1). Stems erect or ascending (growing upward) . . . . (4).
   3. Stems prostrate, creeping, or spreading . . . . (6).

4 (3). Leaves without a terminal bristle, margins sparsely
        toothed . . . . 10. S. *viridissima*.
   4. Leaves with a terminal bristle, margins ciliate . . . . . . (5).

5 (4). Bristle at the apex of leaves about 1 mm long
        . . . . 7. S. *rupincola*.
   5. Bristle at the apex of the leaves shorter than 0.5 mm
        . . . . 4. S. *xneomexicana*.

6 (3). Stems definitely dorsiventral (with distinct upper and
        lower surfaces), underside leaves often larger than those
        of the upper side; leaves appearing strongly upswept
        when viewed from the side . . . . (7).
   6. Stems not dorsiventral, leaves similar in size and
        symmetrically arranged on the stem; leaves not appearing
        strongly upswept when viewed from the side . . . . . . (10).

7 (6). Leaves tapering to a short, stout, yellowish hardened
        leaf tip . . . . 11. S. *wrightii*.
   7. Leaves abruptly (suddenly) acute (with a short, sharp tip)
        at the apex, bristle slender, whitish . . . . (8).

8 (7). Stems short, forming tufted (clumped), cushiony mats;
        rare plants of S Brewster Co.. . . . . 8. S. *scopulorum*.
   8. Stems longer, forming open or compact mats . . . . . . (9).

9 (8). Leaves on the underside of the stems linear to linear-
        lanceolate; bristles usually 0.5 mm long or more;
        sporophylls bristle-tipped . . . . 5. S. *peruviana*.
   9. Leaves on the underside of the stems lanceolate; bristles
        usually 0.3 mm or less long; sporophylls acute to acuminate
        (tapering to a drawn-out point). . . . . 1. S. *arizonica*.

10 (6). Leaves often spreading, linear-lanceolate;
       bristles prominent . . . . 9. *S. underwoodii.*
    10. Leaves closely appressed, ovate-elliptic;
       bristles short or absent . . . . 3. *S. mutica.*

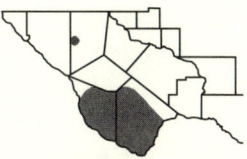

B   1. *Selaginella arizonica* Maxon, ARIZONA
SPIKEMOSS (for the state of AZ). Fig. 5. Typically S-C Trans-Pecos in soil, rock crevices, gravel
areas, on rocky slopes, ledges, igneous rock, sandstone, or limestone, especially the Caballos Novaculite area S of Marathon, Brewster Co.; also Sierra Diablo Mts.,
Culberson Co. 3,500–6,500 ft. C TX, AZ. Mexico.

    Selaginella arizonica is similar to the more common S. *peruviana.* Ari-
zona spikemoss is most easily distinguished by the thin, lanceolate leaves
on the underside of the stem and the lack of bristles on the sporophylls. It
occasionally grows intertwined with S. *rupincola.*

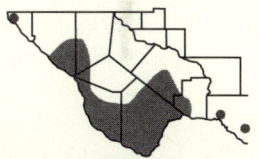

B   2. *Selaginella lepidophylla* (Hook. &
Grev.) Spring, RESURRECTION PLANT, SIEMPRE
VIVA, FLOWER OF STONE (*lepido* = scale +
phyll = leaf). Fig. 5. Often locally common on
dry, rocky slopes, bluffs, and ledges throughout
much of the Trans-Pecos. Usually on limestone, occasionally igneous or
sandstone substrates. 1,800–6,500 ft. W TX, NM. Mexico.

    The rosettes of S. *lepidophylla* are flat and bright green when moist and
may measure to 25 cm across. Because this species occurs in abundance
only in north-facing habitats, it is a good indicator of direction to those
traveling in desert areas where the species is prevalent. The name resurrection plant comes from the way desiccated plants, which roll up into brown,
nestlike balls, will revive, uncurl, and resume normal growth after receiving
water, even after years of being dry. The species is sold as a house plant
and a novelty.

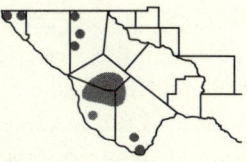

B,G 3. *Selaginella mutica* D. C. Eaton ex
Underw., BLUNT SPIKEMOSS (*muti* = blunt).
Fig. 5. On N- or S-facing slopes, in crevices, on
boulders, cliffs, canyons, usually on igneous rock
or soil but occasionally on limestone or sandstone.
Collections of *Selaginella mutica* var. *mutica,* the more common variety in
the Trans-Pecos, have been made from a variety of mountain ranges in the
Trans-Pecos including the Franklin and Hueco Mts., El Paso Co.;

Guadalupe, Sierra Diablo, and Delaware Mts., Culberson Co.; Davis Mts., Jeff Davis Co.; Sierra Vieja and Chinati Mts., Presidio Co.; Rosillos and Chisos Mts., Brewster Co. 3,950–8,000 ft. W TX, NM, AZ, CO, UT, WY.

Selaginella mutica is most easily distinguished by the ovate-elliptic, boat-shaped leaves, short bristles (less than 0.5 mm long or absent), and the subtruncate apex of the leaves, which resemble a boat prow when viewed in profile. Blunt spikemoss was reported by Tryon (1955) to replace Underwood spikemoss eventually when the two taxa grow in the same mat. Presently two varieties of *S. mutica*, which may be difficult to distinguish, are recognized.

Key to the Varieties

1. Margins of leaves and sporophylls with long, spreading cilia; bristles at the leaf tips short (0.03–0.06 mm long)
   . . . . 3a. *S. mutica* var. *mutica.*
1. Margins of leaves and sporophylls with short, ascending cilia; bristles at the leaf tips longer (0.2–0.45 mm long)
   . . . . 3b. *S. mutica* var. *limitanea.*

3a. *Selaginella mutica* D. C. Eaton ex Underw. var. *mutica.* Two Sul Ross collections of var. *mutica* from Pine Canyon in the Chisos Mts., Brewster Co., show intermixing with *S. underwoodii,* and one collection from the Rosillos Mts., Brewster Co., has *S. peruviana* included in the mat.

3b. *Selaginella mutica* D. C. Eaton ex Underw. var. *limitanea* Weath. Sul Ross collections of var. *limitanea* are mainly from the Mt. Livermore area of the Davis Mts., Jeff Davis Co. where the variety is often found in mixed mats with *S. underwoodii.*

B   4. *Selaginella xneomexicana* Maxon, NEW MEXICO SPIKEMOSS (for the state of NM). Fig. 5. Rare in the Trans-Pecos on boulders, in limestone crevices in the Franklin Mts., El Paso Co.; in igneous soil in Green Gulch, Chisos Mts., and 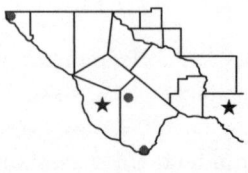 Sunny Glen, Brewster Co.; also reported from Presidio and Val Verde Counties (Correll, 1955). 4,000–5,500 ft. NM, AZ.

This entity is thought to be a hybrid between *S. rupincola* and *S. mutica,* although other possible origins have been proposed. The spores of

S. xneomexicana are abortive (imperfectly developed) and a number of its characters appear to be intermediate between those of the putative parents.

B,G 5. *Selaginella peruviana* (Milde) Hieron.,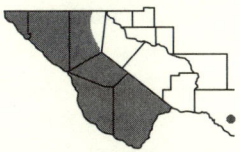
PERUVIAN SPIKEMOSS (for the country of Peru).
Fig. 5. [*S. sheldonii* Maxon]. Locally common
throughout the Trans-Pecos mountains, mesas,
canyons, slopes, and hills, usually on igneous rock
or soil; also found on limestone, especially Caballos Novaculite S of Marathon, Brewster Co. 1,200–7,800 ft. C and W TX, NM, OK. Mexico.
South America.

Although similar in appearance to *S. arizonica*, *S. peruviana* is distinguished by thicker and more linear underside leaves; longer, persistent
bristles on the leaves; and the occurrence of bristles on the sporophylls. Peruvian spikemoss occasionally occurs intertwined in the same mixed mats
with *S. rupincola* in the Caballos Novaculite area S of Marathon and with
S. underwoodii or *S. mutica* in the Davis Mountains.

B,G 6. *Selaginella pilifera* A. Braun, HAIRY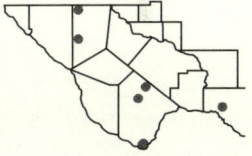
RESURRECTION PLANT, DORADILLA (*pili* =
hair + *fera* = bearing). Fig. 5. Usually in mountainous terrain growing in rock crevices or soil on
slopes, sheltered cliffs, and canyon walls and
floors of the Guadalupe and Sierra Diablo Mts., Culberson Co.; Del
Norte, Glass, and Chisos Mts., Brewster Co.; and Seminole Canyon, Val
Verde Co.; most frequently on limestone, but also occurring in the Chisos
Mts. on igneous substrates. 4,000–8,000 ft. NM. N Mexico.

The long bristles (0.5–1.5 mm) at the apexes of the leaves give this species its "hairy" appearance and allow it to be distinguished easily from the
similar and more common *S. lepidophylla*. *Selaginella pilifera* and *S.
lepidophylla* are the only two Trans-Pecos selaginellas that have the "resurrection" rosette habit.

7. *Selaginella rupincola* Underw., LEDGE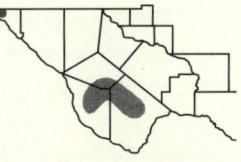
SPIKEMOSS, ROCKLOVING SPIKEMOSS (*rupi* =
rock + *cola* = dwelling). Fig. 5. On cliffs, ledges,
and hillsides in igneous rock areas in El Paso, Jeff
Davis, Presidio, and Brewster Counties, and on
Caballos Novaculite S of Marathon. 3,400–6,800 ft. NM, AZ. Mexico.

The linear leaves of *S. rupincola* are closely and evenly imbricate (overlapping) around the stem, and the sterile branches have a tufted appearance

that results from the thick clusters of long-bristled leaves at the branch tips. Ledge spikemoss is presumed to be a parent of *S. xneomexicana*. It occasionally occurs in the same mat with *S. arizonica* or *S. peruviana*, both of which have dorsiventral stems with upswept leaves.

B   8. *Selaginella scopulorum* Maxon, ROCKY MOUNTAIN SPIKEMOSS (*scopul* = a rock, cliff, or crag). Fig. 5. [*S. densa* var. *scopulorum* (Maxon) R. M. Tryon]. The only location where *S. scopulorum* has been collected in Texas is on 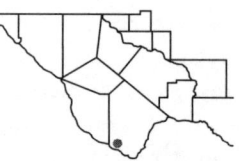 sandstone outcrops 3 mi E of Study Butte, Brewster Co. In other parts of its range this taxon is found from alpine tundra to desert slopes, usually in dry, rocky habitats on igneous or sedimentary substrates. 2,600–2,900 ft. Rocky Mountain axis into Canada, coasts of OR, WA.

   Selaginella scopulorum is a member of the *S. densa* complex, which is in need of further systematic study. Specimens from the Study Butte population are yellow-green in color and have very short, tufted branches that form a flat, cushiony mat.

B   9. *Selaginella   underwoodii*   Hieron., UNDERWOOD SPIKEMOSS (for L. M. Underwood, American fern specialist). Fig. 5. In crevices, on boulders or soil, on rocky slopes, bluffs, and canyons on igneous or limestone substrates,  generally in moist or shaded areas; Franklin Mts., El Paso Co.; Davis Mts., Jeff Davis Co.; Chisos Mts., Brewster Co.; and Seminole Canyon State Historic Park, Val Verde Co.; usually above 5,800 ft. (1,400) 4,000–8,000 ft. W TX, NM, AZ, CO, UT, WY, OK. Mexico.

   Selaginella underwoodii somewhat resembles *S. xneomexicana* but is distinguished by the leaves being more lax and in fewer ranks than in *S. xneomexicana*. Underwood spikemoss often grows in mixed mats with *S. mutica* and occasionally with *S. peruviana*.

B   10. *Selaginella   viridissima*   Weath., SLENDER SPIKEMOSS, GREEN SPIKEMOSS (*virid* = green + *issima* = -est). Fig. 5. [*S. coryi* Weath.]. In the United States, this species is most likely to be found growing on large boulders, 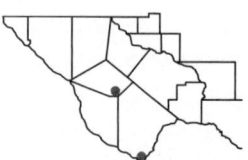 ledges, canyon walls, and mountain slopes in the Chisos Mts., Brewster Co. Slender spikemoss is rarely found in the Davis Mts., Jeff Davis Co. (1,400) 4,500–7,500 ft. Coahuila, Mexico.

Elevations not typical for the species appear in parentheses.

Selaginella viridissima and *S. mutica* are similar in that both may produce leaves without a terminal bristle. *Selaginella viridissima* is distinguished by an erect habit and linear-lanceolate leaves with dentate (toothed) margins, while *S. mutica* exhibits a creeping habit and elliptic leaves with ciliate margins.

B,G 11. *Selaginella wrightii* Hieron., WRIGHT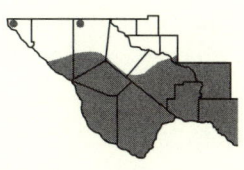
SPIKEMOSS (for C. W. Wright, noted botanical
collector). Fig. 5. Throughout much of the Trans-
Pecos on limestone boulders, ledges, rocks, and
soils, on mesas, canyons, creek banks, and hills.
1,000–6,000 ft. W TX, NM. Mexico.

Wright spikemoss is low, creeping, and mat-forming, often producing abundant, erect strobili (sporebearing cones), which arise from the spreading mat. The short, stout bristle at the leaf tip may be poorly differentiated from the leaf tissue and may appear as a hardened yellowish leaf apex.

2. **Equisetaceae** Michx. ex DC., Horsetail, Scouring Rush Family

These rushlike terrestrial plants with perennial rhizomes grow in moist places. Their cylindrical, jointed, and longitudinally (lengthwise) ridged aerial (aboveground) stems are either annual or perennial and have distinct, solid nodes and hollow internodes. Branches are present at the nodes of some species. The leaves are reduced, whorled, and united to form a sheath with persistent or deciduous apical teeth (at the top of the sheath). The cones are an aggregation of stalked, peltate, scalelike sporophylls that terminate vegetative stems or specialized brown reproductive stems in some species. The tips of the cones may be rounded or sharply pointed. The sporangia number 5–10 per sporophyll, are pendant (hang downward) from the underside, and dehisce (split open) longitudinally. The spores, which are of one kind, are numerous, green (white in hybrids), and equipped with four tendril-like appendages. These appendages, called elaters, coil when moist and uncoil when dry and presumably play a role in spore dispersal. One can readily observe the coiling of the elaters by gently breathing on a fertile cone, while looking through a magnifying lens. Gametophytes are about 25 mm in diameter, unisexual, terrestrial, and green.

The fossil record shows little evolutionary change in members of Equisetaceae since giant horsetails were a dominant part of the Carboniferous period flora in the Paleozoic era 300 million years ago. The name

Equisetum comes from the Latin *equus* for horse and *seta* for bristle, referring to the coarse, wiry, hairlike roots of some species. The name is also attributed to the resemblance of the whorled branches on the stem of *E. arvense* to that of a fluffy horsetail. The family contains only one genus. Distribution of the 15 species is nearly worldwide. Approximately 11 species occur in the United States, with two species and one hybrid occurring in the Trans-Pecos. One additional Texas species, the above-mentioned *E. arvense,* is reported from a single location in Lubbock County.

1. *EQUISETUM* L., HORSETAIL, SCOURING RUSH

The characters of the genus are the same as those of the family. Recent studies of hybridization among the species of *Equisetum* that are reported to occur in the Trans-Pecos have clarified previously existing problems concerning intergrading and confusing characters. Because members of *Equisetum* grow in moist habitats, recent droughts in the area may have reduced scouring rush localities in the Trans-Pecos.

The stems of some species of *Equisetum* contain deposits of minute silica particles in the epidermal cells. This characteristic made them useful for cleaning pots and pans in colonial and frontier times and earned them the name "scouring rushes." In addition to scouring and polishing, *Equisetum* has been used for a variety of internal and external medicinal purposes, cosmetically as a hair and nail strengthener, for weaving mats, and for making a yellow ochre dye. Although seventh-century Romans ate young *Equisetum* shoots like asparagus, members of the genus are reported to be toxic to livestock and questionable for human consumption. Ingesting large quantities of *Equisetum* is reported to disturb thiamine metabolism.

Key to the Species

1. Sheath with 1 dark band (at the tip); cone apex rounded
   to bluntly apiculate (with a small, slender, pointed tip);
   spores green . . . . 3. *E. laevigatum.*
1. Sheath with 2 dark bands (1 in middle, 1 at tip);
   cone apex sharply pointed; spores green or white. . . . . (2).

2 (1). Spores green, spherical. . . . . . . 2. *E. hyemale* ssp. *affine.*
   2. Spores white, misshapen . . . . 1. *E. xferrissii.*

1. *Equisetum xferrissii* Clute, FERRISS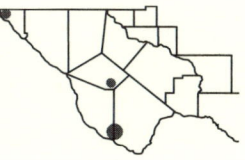
SCOURING RUSH (for J. H. Ferriss, fern bota-
nist). Moist areas along streams and rivers in rare,
scattered locations in the Franklin Mts., El Paso
Co.; Davis Mts., Jeff Davis Co.; and Bofecillos
Mts., Presidio and Brewster counties. 2,800–5,300 ft. Throughout
much of the United States (except the deep South), into S Canada. N
Mexico.

Ferriss scouring rush is known to be a hybrid between *E. laevigatum* and
E. hyemale ssp. *affine*. *Equisetum xferrissii* reproduces vegetatively. Mature cones may be produced, but the misshapen white spores are sterile and
are not shed.

2. *Equisetum hyemale* ssp. *affine* (Engelm.)
Calder & R. L. Taylor, COMMON SCOURING
RUSH, CANUELA (*hyem* = winter, referring to
the perennial nature of the stems). Fig. 6. [*E.*
hyemale var. *robustum* (A. Braun) A. A. Eaton;
E. prealtum Raf.; *E. robustum* A. Braun]. Although common scouring
rush has been reported to occur in the Trans-Pecos, its presence in the area
is uncertain. 1,000–4,000 ft.

The gray-green stems of this species may grow to be over 2 m tall and,
being perennial, persist for more than one season. Common scouring rush
is also called tall or great scouring rush and occurs throughout the United
States, much of Canada, Mexico, and Guatamala (in Central America).

B,G   3. *Equisetum laevigatum* A. Braun,
SMOOTH SCOURING RUSH, AOTA (*laevigat* =
smooth, slippery). Fig. 6. [*E. funstonii* A. A.
Eaton; *E. kansanum* J. H. Schaffn.]. Along
streams, creeks, and wet places in McKittrick
Canyon, Guadalupe Mts., Culberson Co.; Davis Mts., Jeff Davis Co.; the
lower canyons of the Rio Grande, Brewster Co.; and Big Bend Ranch
State Park, Brewster and Presidio Counties; reported but unconfirmed in
Big Bend National Park, Brewster Co. (National Park Service, 1995).
(1,700) 4,800–6,800 ft. Throughout much of the Midwest and west into
S Canada. N Mexico.

The name *Equisetum laevigatum* was applied in the past by some taxonomists to the entity we now know to be the hybrid *E. xferrissii*. The name *E.*
kansanum was previously used for the plants we now recognize to be the
true *E. laevigatum*. Smooth scouring rush has yellow-green stems and

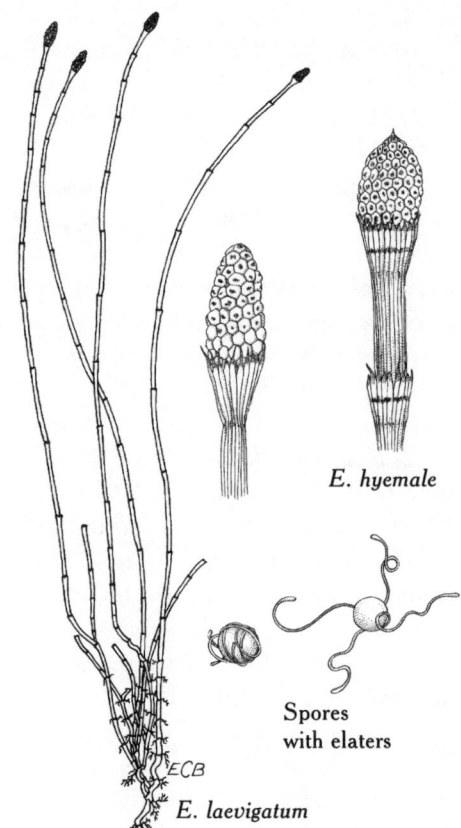

E. hyemale

Spores
with elaters

Fig. 6. *Equisetum laevigatum*, habit;
cone. *E. hyemale*, cone. Spores with
elaters.

E. laevigatum

sheaths that usually lack a central, dark girdling band, although there may be occasional, obscure girdling on some sheaths toward the base of the stem. The stems are seldom over 1.2 m tall. The cone apexes are rounded to apiculate with blunt tips. In contrast, both common and Ferriss scouring rushes generally have dark-girdled sheaths and pointed cone apexes. The Trans-Pecos populations of *E. laevigatum* generally retain their stems over winter. In colder climates, the stems of this species die to the ground each winter.

3. **Ophioglossaceae** C. Agardh, Adder's Tongue Family

Plants of the adder's tongue family are perennial and usually terrestrial, although a few are epiphytic. The stems are unbranched, simple, and upright with clasping leaf bases that form an open or closed sheath. Leaves number one or two per stem, and the common stalk gives rise to a sterile blade, called a trophophore, which is entire or variously compound, and to one or more erect, fertile, spore-bearing spikes or panicles, called sporophores. The naked sporangia occur in two rows and open by transverse (crosswise) slits. The numerous yellow spores are all of one kind, and the subterranean gametophytes are usually fleshy and not green.

The name Ophioglossaceae is from the ancient Greek *ophis* for serpent, and *glossa* for tongue, alluding to the narrow fruiting spike. The family has nearly worldwide distribution and contains five genera and seventy to eighty species. Three genera and about thirty-six species occur in the United States, two genera and nine species occur in Texas, and one genus and two species are found in the Trans-Pecos.

1. *OPHIOGLOSSUM* L., ADDER'S TONGUE

Ophioglossums are terrestrial, small, somewhat fleshy plants with fibrous roots and either a short or a bulbous caudex that forms a subterranean stem. The trophophore blades are glabrous (lacking hairs or scales), simple, and entire, with reticulate venation (netlike veins). Trophophore blades per stem may number one or more. The sporophores are slender, erect, stalked, unbranched, and arise from the stipe. Sporophores may be one per stem or absent, and the yellow sporangia are two ranked.

Several species of *Ophioglossum* have high chromosome numbers, and one species, *O. reticulatum* L. ($2n$ = ca. 1440), has what is perhaps the highest chromosome number in the plant kingdom (Stace, 2000).

Adder's tongue favors open, disturbed, grassy areas and often appears after rains. These small plants, with their resemblance to the early leaves of some monocots, may easily be overlooked. Twenty-five to thirty species are distributed nearly worldwide, mainly in tropical and subtropical areas. Five species occur in Texas, with two species occurring in the Trans-Pecos and adjacent areas.

Key to the Species

1. Roots more than 8 per stem; trophophore blades
   usually apiculate, veins forming smaller areoles within
   large areoles (fig. 7); Brewster, Jeff Davis, and
   Presidio counties . . . . 2. *O. polyphyllum.*
1. Roots 8 or fewer per stem; trophophore blades usually
   acute, not apiculate, veins forming areoles with branched
   or unbranched, nonareole fragments within; in sand
   dunes, Winkler Co. . . . . 1. *O. petiolatum.*

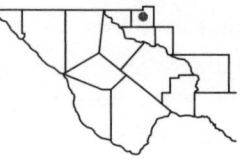

1. *Ophioglossum petiolatum* Hook., STALKED ADDER'S TONGUE, OLD WORLD ADDER'S TONGUE (*petiol* = stalk). Fig. 7. One collection, "locally abundant in moist sand dunes" NE of Kermit in Winkler Co., 29 May 1949. 2,800–2,900 ft. S U.S., W to TX. West Indies. Mexico. N South America. Asia. Pacific Islands.

The single collection of *O. petiolatum* mentioned was made by Rogers McVaugh. The taxon has not been recollected in the sand dune country or elsewhere in western Texas. Stalked adder's tongue is an easily cultivated plant, growing readily in pots, and may become a greenhouse weed. This species was probably introduced to North America in the early 1900s. The first collections on the continent were made between 1900 and 1930.

Both of the Trans-Pecos adder's tongues form large colonies when favorable growing conditions occur. *Ophioglossum polyphyllum* and *O. petiolatum* have leaf blades of similar shape (oblong-elliptic to lanceolate) and overlapping size, although the leaves of *O. petiolatum* tend to be smaller. The areoles of the trophophore veins can be difficult to see, even with a magnifying lens. The best characters for distinguishing the Trans-Pecos species of *Ophioglossum* may be the number of roots per stem, the leaf apex, and the area of occurrence.

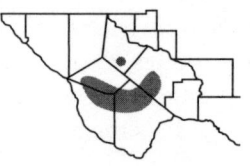

2. *Ophioglossum polyphyllum* A. Braun, CHIHUAHUAN DESERT ADDER'S TONGUE (*poly* = many + *phyll* = leaf). Fig. 7. Sparse or locally abundant at roadsides among grasses in clay soil, often in limestone, also in igneous regions in Jeff Davis, Presidio, Brewster, Pecos, and Reeves Counties; often appearing after rain. (3,500) 4,000–4,700 ft. TX, AZ, HI. Mexico. Africa. Asia.

The first collections of Chihuahuan desert adder's tongue from the Trans-Pecos were made in 1995. Initially these collections were thought to be *O. engelmannii*, which occurs across the southern United States and extends into eastern Texas and Arizona. Currently the Trans-Pecos collections are recognized as *O. polyphyllum*, which is widespread in the Old World but had not been previously reported from North America (Zech et al., 1998). The typical habitat of *O. polyphyllum* is in shallow roadside ditches and depressions in typical plains grassland. The leaves appear aboveground after heavy rains and are evident only for a period of about two weeks.

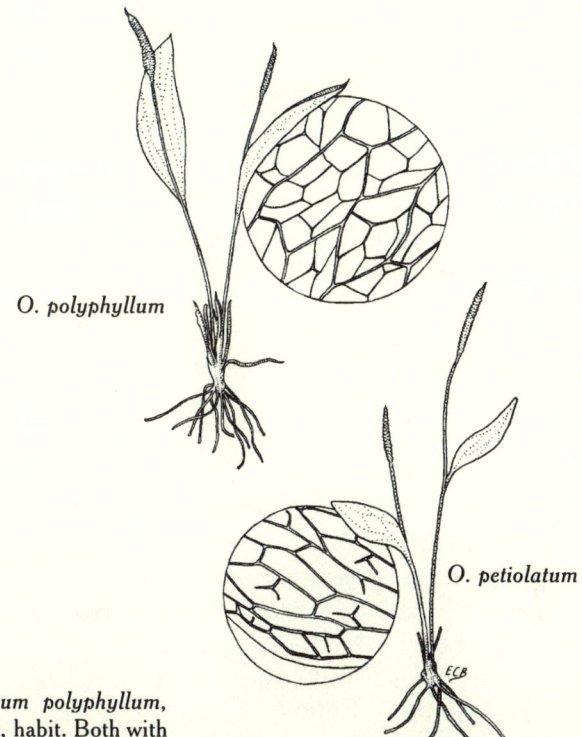

O. polyphyllum

O. petiolatum

Fig. 7. *Ophioglossum polyphyllum*, habit. *O. petiolatum*, habit. Both with enlarged view of vein areoles.

4. **Anemiaceae** Link, Anemia Family

Members of the Anemia family are terrestrial or rock-dwelling perennials. The stems are compact or short-creeping, horizontal, hairy rhizomes. The leaves are usually erect (rarely a flat rosette) and partially to completely dimorphic (taking two forms). The sessile (without a stalk), oblong sporangia, which open apically, occur in two rows at the distal (tip) ends of the fertile segments. Gametophytes are terrestrial, cordate (heart-shaped), and green, with unequal lobes. The family consists of two genera that are widespread in tropical and subtropical regions of the world. One genus and three species occur in the United States.

1. *ANEMIA* Sw., ANEMIA

Members of the genus *Anemia* have rhizomatous, horizontal, short-creeping stems with dark hairs. The leaves are either fully dimorphic, with the blade and fertile parts separate, or partially dimorphic, with sporangia at the ends of erect, divided, spikelike fertile pinnae that arise from the petiole just at the base of the sterile blade section. The blades are 1–3-pinnate and leathery or papery in texture. The approximately 117 species are distributed mainly in tropical and subtropical regions, with the greatest number of species occurring in Brazil and Mexico.

The three species found in the United States usually grow on limestone, two species being endemic to Florida and one to Texas. The name *Anemia* is from the Greek word *aneimon,* meaning without clothing, referring to the naked, unprotected sporangia.

B   1. *Anemia mexicana* Klotzsch, MEXICAN ANEMIA, CHALK LEDGE FERN (for Mexico). Fig. 8. A collection of *Anemia mexicana,* reported as "infrequent in limestone crevices," was made at the mouth of the Pecos River, Val Verde 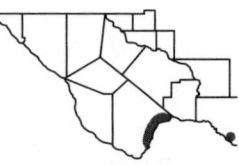 Co. in 1948; the site has changed greatly since that time, and the continued occurrence of this population is questionable; *Anemia mexicana* has also been found growing near seeps and springs in a few locations in the lower canyons of the Rio Grande, Brewster Co.; it is reported but unconfirmed in Big Bend National Park, Brewster Co. (National Park Service, 1995); the species is frequently found on lightly shaded limestone outcrops on the Edwards Plateau. 1,000–1,600 ft. TX. Mexico.

The fronds of *A. mexicana* may grow to be 45 cm long, and the somewhat leathery, once-pinnate blades are divided into 4–7 pairs of finely toothed pinnae, which are usually alternate (borne at different levels on the stem). The two spikelike fertile pinnae, which arise just below the base of the sterile blade section, often surpass the sterile section in height.

Chalk ledge fern is becoming popular for landscaping and serves as an attractive ground cover in suitable locales. The species has been used by indigenous peoples in Mexico for various medicinal purposes.

5. **Pteridaceae** Rchb., Maidenhair Fern Family

Members of the Pteridaceae are usually perennial, occasionally annual, and grow on rock or in soil. The stems are rhizomatous, compact to creeping, with scales and/or hairs. The petioles generally have some scales on their lower portions. The leaves are monomorphic (of one type) or dimorphic, some species having circinate buds (coiled at the tip), some not; blades are 1–6-pinnate and glabrous or variously clothed with hairs, glands, scales, and/or farina. Sori occur toward the ends of the veins, often forming a band near the margin, or covering the undersurface. False indusia (fig. 2), when present, are formed by reflexed leaflet margins. Gametophytes are above-ground, green, and reniform (kidney-shaped) to obcordate.

There has been much controversy concerning the composition, individ-ual taxa, and even the name of this family of ferns. Recently the name Adiantaceae has been used, and for many years the taxa were included in the Polypodiaceae. The treatment here is from Windham (1993a), in which determinations are based not only on morphological characters but also take into consideration chromo-somal and biochemical information. Basic morphological characters used to differentiate the family include sori that either lack indusia or have false indusia formed by revolute or reflexed leaflet margins.

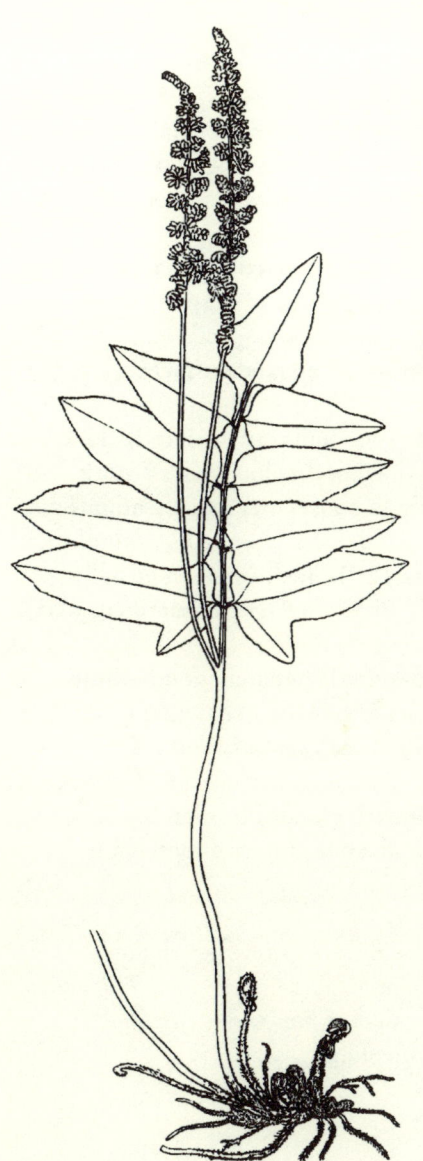

About 40 genera and 1,000 spe-cies are found worldwide, with 13 genera and approximately 90 spe-cies occurring in the United States. Seven genera and 41 species occur in Texas, and seven genera with 38 species are found in the Trans-Pecos.

Fig. 8. *Anemia mexicana,* habit.

Key to the Genera

1. Sporangia occurring on the actual surface of the recurved marginal lobes of the pinnules (leaflets) . . . . 1. *Adiantum.*
1. Sporangia not occurring on the recurved marginal lobes but rather on the lower pinnule surface itself, which may curl over to form a false indusium that covers the sporangia . . . . (2).

2 (1). Blades once-pinnate to pinnate-pinnatifid, lower surface densely covered with ciliate scales. . . . . 3. *Astrolepis.*
   2. Blades 2–6-pinnate or, if less divided, then abaxial (lower) surfaces glabrous (without hairs) or pubescent (with hairs), lower surface not densely covered with ciliate scales . . . (3).

3 (2). Pinnae with white or yellow farina on abaxial surfaces (concealed beneath fringed or stellate (star-shaped) scales in *N. aschenborniana*) . . . . (4).
   3. Pinnae completely lacking farina on abaxial surfaces . . . (5).

4 (3). Rhizome scales either concolored (uniformly colored) dark brown to black or bicolored (having 2 colors) with a dark central stripe; pinnules sessile to subsessile, usually adnate (fused) to the midrib . . . . 6. *Notholaena.*
   4. Rhizome scales concolored tan or brown; pinnules usually distinctly stalked . . . . 2. *Argyrochosma* (in part).

5 (3). Blades clearly pentagonal (5-sided), pedately divided into 3 deeply pinnate-pinnatifid segments (fig. 12); abaxial pinnae pubescence consisting of scales and coiled and straight hairs . . . . 4. *Bommeria.*
   5. Blades 2–4-pinnate or, if pinnate-pinnatifid, then blades linear to lanceolate in shape; abaxial pinnae pubescence various but not as above . . . . (6).

6 (5). Largest pinnules usually more than 4 mm wide; rhizome scales generally strongly bicolored. . . . . 7. *Pellaea.*
   6. Largest pinnules usually less than 4 mm wide; rhizome scales concolored to weakly bicolored. . . . . (7).

7 (6). Pinnules glabrous, somewhat cordate (heart-shaped) at the
base, attached to lustrous (glossy), dark-colored stalks
. . . . 2. *Argyrochosma* (in part).
7. Pinnules variously pubescent or glabrous, mostly rounded,
truncate (straight across), or cuneate (wedge-shaped) at
base, sessile or attached to variously colored stalks
. . . . 5. *Cheilanthes.*

1. *ADIANTUM* L., MAIDENHAIR FERN

Trans-Pecos ferns of the genus *Adiantum* grow to be 10–60 cm tall and
prefer cool, moist habitats near streams, waterfalls, and springs. The stems
are rhizomatous with golden to medium brown scales, which continue on
the lower petioles. The slightly dimorphic leaves are pendant to ascending
with glabrous blades that may become four or more times pinnate. The numerous fan-shaped to rhombic (diamond-shaped) pinnules are about as
broad as they are long, with veins that are dichotomously branched
(forked) near the base and parallel toward the distal ends. The sporangia
are unusual in that they are submarginal on the abaxial surface of the false
indusia that are formed by reflexed segment margins, rather than occurring
on the underleaf surfaces themselves (fig. 9).

The name *Adiantum* is from the Greek *adiantos*, meaning unwetted, referring to the glabrous leaves that readily shed water drops. The widely distributed species of *Adiantum* have been used for various medicinal purposes, especially hair and scalp conditions, since ancient times. Many species of maidenhair ferns are cultivated for their delicate beauty. Although
the 150 to 200 species occur nearly worldwide, the genus is mainly
tropical.

Nine species of *Adiantum* are found in the United States, primarily in
moist habitats. Two species occur in Texas, one being found in the
Trans-Pecos.

B,G 1. *Adiantum capillus-veneris* L., MAIDEN-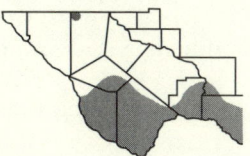
HAIR FERN, VENUSHAIR FERN (*capill* = hair +
vener = pertaining to Venus). Fig. 9. Moist canyons, cliffs, and bluffs, near seeps, waterfalls, and
springs in cool protected locales on limestone or
igneous substrates in Culberson, Jeff Davis, Presidio, Brewster, Terrell,
and Val Verde Counties. 1,000–6,200 ft. The Panhandle, E and S-C
TX, E to VA, SC, FL; W to CA; N to CO, UT, MO, SD. British

Fig. 9. *Adiantum capillus-veneris*, habit;
pinnule.

Columbia, Canada. Mexico. West Indies. Central America. South America. Eurasia. Africa.

Maidenhair fern is best distinguished by the unique flaplike false indusia that fold over flat against the lower leaflet surface, thus covering and protecting the sporangia that occur on the surface of the recurved marginal lobes between two leaflike layers. Pecos River collections previously identified as *A. tricholepis* are now identified as *A. capillus-veneris*.

2. *ARGYROCHOSMA* (J. Smith) Windham, SILVER FERN, FALSE CLOAK FERN

Members of the genus *Argyrochosma* can grow to be 30 cm tall, but they are generally smaller and delicate in appearance. Silver ferns usually are found growing on rock. The stems are rhizomatous, compact, and short-creeping, with scales that are thin, linear to lanceolate in shape, and concolored, tan or red-brown. The stipes are dark, either brown or black, and glabrous or with a few scales or hairs toward the base. The leaf blades are ovate to lanceolate, 2–6-pinnate, with pinnules that are glabrous or silvery farinose beneath. The pinnules are generally less than 4 mm wide and often distinctly stalked. The false indusia are marginal and generally poorly defined. The sporangia may be marginal or submarginal or occur along the veins.

The name *Argyrochosma* is from the Greek *argyros*, meaning silver, and *chosma* for powder, referring to the silvery farina that covers the lower pinnule surface of some species. The taxa currently assigned to the genus *Argyrochosma* have been variously included in *Pellaea*, *Cheilanthes*, and *Notholaena*. Recent morphological, chemical, and chromosomal studies by Windham (1987) have suggested a close relationship between *Argyrochosma* and *Pellaea*. *Argyrochosma* is most easily distinguished from *Pellaea* by the combination of concolored rhizome scales and pinnules that are less than 4 mm wide. The chromosome number of *Argyrochosma* ($x = 27$) is unique for this type of cheilanthoid fern. About 20 species occur in North America, Mexico, Central and South America, and the West Indies. Six species are found in the United States, three in Texas, and two in the Trans-Pecos.

Key to the Species

1. Undersurface of the pinnules lacking silver farina; pinnules articulate (jointed to the stem), dark color of stalks stopping at pinnule bases . . . . 2. *A. microphylla.*
1. Undersurface of the pinnules densely covered with silver farina; pinnules not articulate, dark color of stalks continuing into the pinnule bases abaxially 1. *A. limitanea* ssp. *mexicana.*

B,G 1. *Argyrochosma limitanea* (Maxon) Windham ssp. *mexicana* (Maxon) Windham, MEXICAN SILVER FERN, SOUTHWESTERN FALSE CLOAK FERN (*limit* = border, bounded, probably referring to distribution along the  Mexico-U.S. border). Fig. 10. [*Notholaena limitanea* Maxon ssp. *mexicana* Maxon; *N. limitanea* var. *mexicana* (Maxon) M. Broun]. Rare on limestone cliffs, Franklin Mts., El Paso Co.; Guadalupe and Sierra Diablo Mts., Culberson Co.; and Chisos and Dead Horse Mts., Brewster Co. 3,000–7,000 ft. NM, AZ. Mexico.

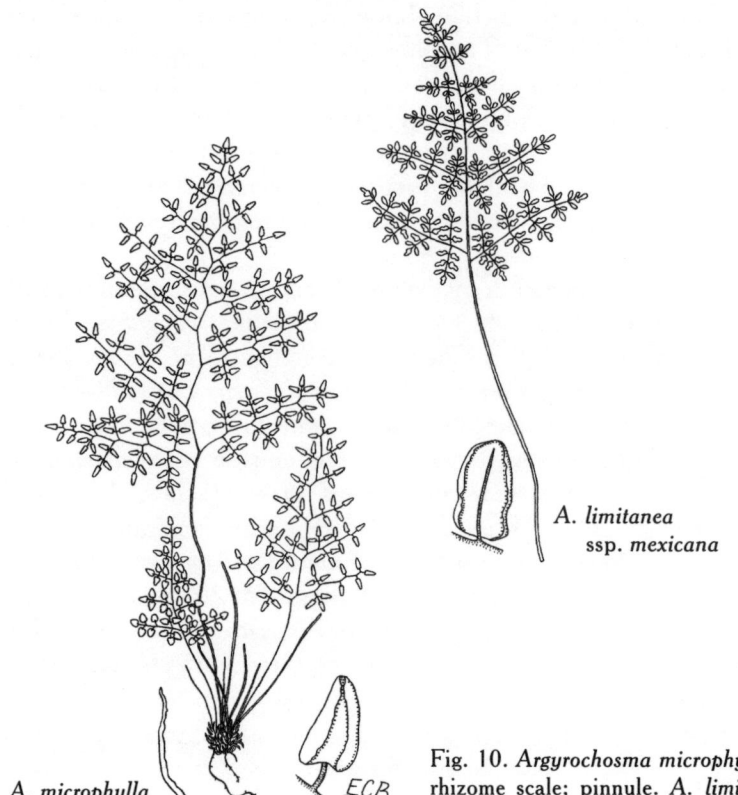

A. *limitanea*
ssp. *mexicana*

A. *microphylla*

Fig. 10. *Argyrochosma microphylla*, habit; rhizome scale; pinnule. *A. limitanea* ssp. *mexicana*, frond; pinnule.

B,G   2. *Argyrochosma microphylla* (Mett. ex Kuhn) Windham, SMALL LEAF SILVER FERN, SMALL LEAF FALSE CLOAK FERN (*micro* = small + *phylla* = leaf). Fig. 10. [*Cheilanthes parvifolia* (R. M. Tryon) Mickel; *Notholaena* 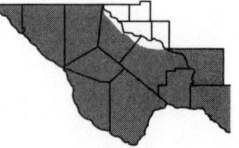 *parvifolia* R. M. Tryon; *Pellaea microphylla* Mett. ex Kuhn]. Frequent to infrequent, usually on limestone substrates in canyons, along streams, or on cliffs, ledges, rocky slopes, and summits, throughout most of the Trans-Pecos. 990–7,500 ft. C TX, NM. Mexico.

Small leaf silver fern differs from the other silver ferns in its lack of silver farina on the underside of the pinnules. The blades of *A. microphylla* are 3–4-pinnate and the pinnules are glabrous, small, distant, numerous, and oval to oblong with cordate bases. This species is reportedly poisonous to sheep.

3. *ASTROLEPIS* D. M. Benham & Windham, STARSCALED CLOAK FERN, SCALY CLOAK FERN

Members of the genus *Astrolepis* may grow to be 130 cm tall, although they are usually considerably smaller. They generally are found on rock but sometimes grow in soil. The stems are rhizomatous with tan to brown linear attenuate scales (tapering at the tip) that may continue up the petiole. The leaf blades are once-pinnate to pinnate-pinnatifid and linear to linear-oblong. The pinnae are ovate, oblong, or long-deltate in shape, with the top surface sparsely to densely covered with stellate or coarsely ciliate scales. The pinnae bottom surface has an underlayer of stellate scales and a top layer of overlapping, ovate to lanceolate ciliate scales; the margins are entire to sinuate-dentate, without false indusia. Sporangia occur along veins near the pinnae margins and often appear as a broad continuous band. Sporulation takes place from summer to fall.

Members of *Astrolepis* have been variously placed in the genera *Notholaena*, *Cheilanthes*, and *Acrostichum*. Morphological and chromosomal studies suggest that the starscaled cloak ferns should be recognized as a separate genus that can be distinguished by the once-pinnate to pinnate-pinnatifid leaf blade, stellate-pectinate (star-shaped, deeply divided) scales on the upper surface of young pinnae, two vascular bundles in the petiole that remain separate below the blade, and a base chromosome number of $x = 29$ (*Notholaena* and *Cheilanthes* are both $x = 30$; Benham and Windham, 1993). The name *Astrolepis* is from the Greek *astro*, meaning star, and *lepis* for scale, referring to the star-shaped scales on the upper surface of the pinnae. Eight species are distributed in the United States, Mexico, Central and South America, and the West Indies. All four U.S. species are found in Texas and the Trans-Pecos.

Key to the Species

1. Largest pinnae generally less than 7 mm long; scales on the upper surface mostly peltate, circular to elliptic, ciliate; scales on the lower surface usually 1 mm or less long, ovate to ovate-lanceolate, obtuse (blunt at the apex) . . . . 1. *A. cochisensis*.
1. Largest pinnae generally more than 7 mm long; scales on the upper surface basifixed, elongate, ciliate; scales on the lower surface usually more than 1 mm long, lanceolate, attenuate . . . . (2).

2 (1). Largest pinnae entire to asymmetrically lobed; scales on
    the upper surface of young pinnae generally dense, the
    body of the scale noticeably wider (5–6 cells wide) than
    the cilia. . . . . 2. *A. integerrima.*
  2. Largest pinnae usually symmetrically lobed; scales on the
    upper surface of young pinnae sparse, the body of the scale
    not much wider (1–4 cells wide) than the cilia . . . . . . (3).

3 (2). Pinnae lobes generally deep and acute; upper pinnae surface
    glabrous to sparsely scaly, the body of most scales about the
    same width as the cilia . . . . 3. *A. sinuata.*
  3. Pinnae lobes generally shallow and broadly rounded; upper
    pinnae surface sparsely scaly, the body of most scales
    discernibly wider than the cilia . . . . 4. *A. windhamii.*

B,G 1. *Astrolepis cochisensis* (Goodd.) D. M.
Benham & Windham, JIMMYFERN, COCHISE
SCALY CLOAK FERN (for Cochise Co., AZ).
Fig. 11. [*Cheilanthes cochisensis* (Goodd.)
Mickel; *C. sinuata* var. *cochisensis* (Goodd.)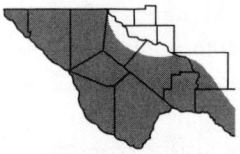
Munz; *Notholaena cochisensis* Goodd.; *N. sinuata* var. *cochisensis*
(Goodd.) Weath.]. Rare to abundant on rocky slopes, bluffs, hills, and
canyons, usually on limestone but sometimes on igneous substrates,
throughout most of the Trans-Pecos including El Paso, Hudspeth,
Culberson, Jeff Davis, Presidio, Brewster, Pecos, Terrell and Val Verde
Counties. 1,200–6,500 ft. Also in plains country and Edwards Plateau.
TX, NM, AZ, CA, OK. Mexico.

    Benham (1992) separated three subspecies of *A. cochisensis* (which are
not morphologically distinguishable) by chromosome number, spore number, and spore diameter. The subspecies *chihuahuensis* is found throughout the Trans-Pecos, while the subspecies *cochisensis* of Arizona, New
Mexico, and California is reported to occur in Texas only in the El Paso
area. *Astrolepis cochisensis* is most easily distinguished from the other
starscaled cloak ferns by the tiny entire or irregularly lobed pinnae and the
peltate scales on the upper pinnae surface. The rhizome scales of *A.*
cochisensis usually have entire margins, while the stem scales of the other
species in *Astrolepis* are ciliate-dentate. Jimmyfern is toxic to sheep and
goats, causing a sometimes fatal nervous syndrome called the "jimmies"
to occur in the animals when they are exercised. The other species of
Astrolepis do not seem to be poisonous to sheep and goats.

B,G 2. *Astrolepis integerrima* (Hook.) D. M. Benham & Windham, WHOLELEAF CLOAK FERN, HYBRID CLOAK FERN (*integer* = whole). Fig. 11. [*Cheilanthes integerrima* (Hook.) Mickel; *Notholaena integerrima* (Hook.) Hevly; *N. sinuata* 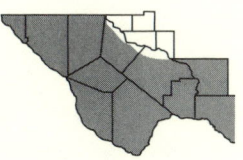 var. *integerrima* Hook.]. Infrequent, usually on limestone, occasionally in igneous soil in mountains, canyons, and hills throughout most of the Trans-Pecos. 990–6,500 ft. W TX, NM, AZ, NV, OK. Mexico.

Characters of *A. integerrima* appear to be intermediate between *A. cochisensis* and *A. sinuata*, leading some botanists in the past to believe that *A. integerrima* is a hybrid between these two species. However, recent isozyme (enzyme) analyses suggest that the putative parents are *A. cochisensis* and an unnamed Mexican taxon. *Astrolepis integerrima* is differentiated from *A. cochisensis* by larger, nearly entire pinnae and abundant, usually basifixed scales on the upper pinnae surface. The asymmetrical lobing on the pinnae as well as the abundance of scales on the upper surface help to differentiate *A. integerrima* from the regularly lobed and sparsely scaled *A. windhamii*.

B,G 3. *Astrolepis sinuata* (Lag. ex Sw.) D. M. Benham & Windham, WAVYLEAF CLOAK FERN (*sinus* = curve, hollow). Fig. 11. [*Acrostichum sinuatum* Lag. ex Sw.; *Cheilanthes sinuata* (Lag. ex Sw.) Domin; *Notholaena sinuata* (Lag. ex 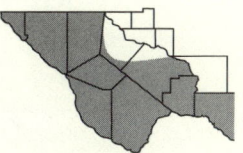 Sw.) Kaulf.]. Infrequent to frequent among rocks and boulders in canyons and mountains throughout much of the Trans-Pecos; most often on limestone, also occasionally on igneous substrates. 2,200–7,200 ft. C and W TX, NM, AZ, GA. Mexico. West Indies. Central America.

Two morphologically similar subspecies of *A. sinuata* are separated by Benham (1992) on the basis of chromosome number, spore number, and spore size. The subspecies *sinuata* is the more common and occurs throughout the entire range of the species, while the subspecies *mexicana* occurs only in the Chisos and Davis mountains of the Trans-Pecos and in SE New Mexico. The symmetrical, deep lobes on the margins of the pinnae of *A. sinuata* are distinctive. By contrast, the lobes on the margins of *A. windhamii* are shallow and rounded.

B,G 4. *Astrolepis windhamii* D. M. Benham, WINDHAM CLOAK FERN (for M. D. Windham, noted fern botanist). Fig. 11. Benham and Windham (1993) show the occurrence of this taxon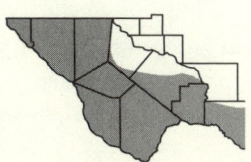

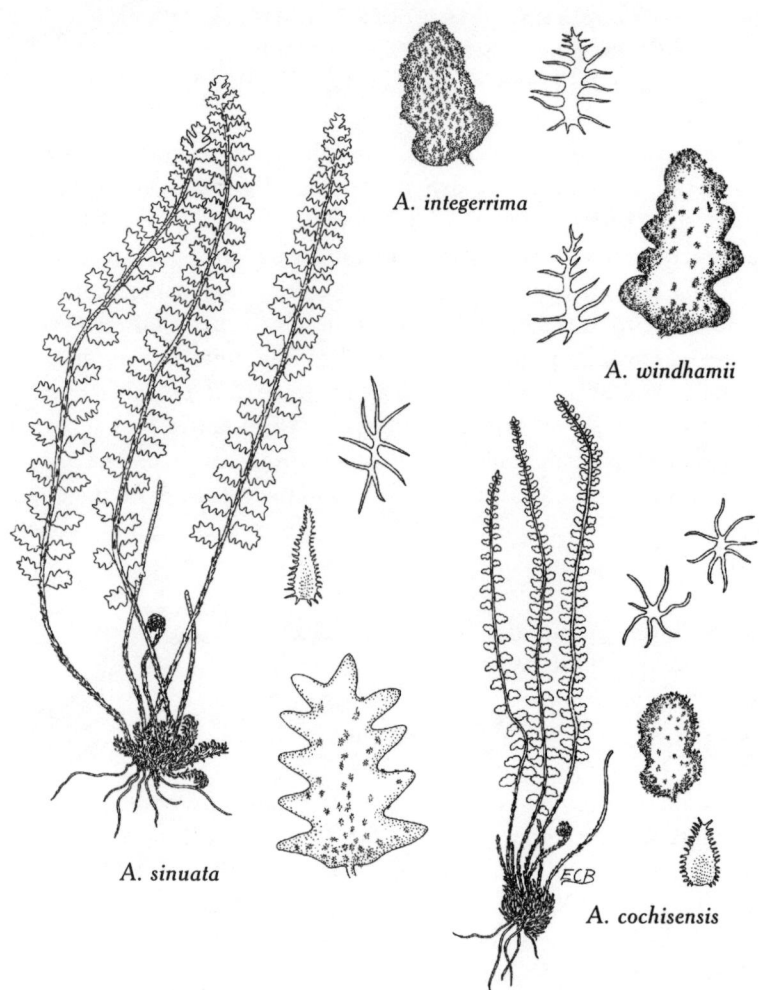

Fig. 11. *Astrolepis sinuata*, habit; pinnule; lower scale; upper scale. *A. integerrima*, pinnule; upper scale. *A. windhamii*, pinnule; upper scale. *A. cochisensis*, habit; pinnule; upper scales; lower scale.

throughout most of the Trans-Pecos on rocky hills and cliffs on limestone and igneous substrates. 1,800–5,500 ft. NM, AZ. Mexico.

Windham cloak fern is a recently described species (Benham, 1992) that is thought to contain a genome from each of three sources: *Astrolepis sinuata*, *A. cochisensis*, and an undiscovered taxon. *Astrolepis windhamii* is best distinguished by symmetrical, crenately lobed (scalloped) pinnae.

The pinnae of *A. sinuata* are more sharply lobed, and those of *A. integerrima* and *A. cochisensis* are asymmetrically lobed or entire.

4. *BOMMERIA* E. Fourn. ex Baill., BOMMERIA

Plants of the genus *Bommeria* are terrestrial and have prostrate, rhizomatous, long-creeping, scaly stems. The stipes are hairy, somewhat scaly, and considerably longer than the blades. The pentagonal leaf blades are pedately divided into three deeply pinnate-pinnatifid segments. The upper pinnae surfaces are covered with needlelike hairs, and the lower pinnae surfaces have scales and both coiled and straight hairs. The sporangia are borne along the veins and margins of the pinnae, and there are no false indusia.

The genus name *Bommeria* honors J. E. Bommer, a Belgian botanist. Five species of *Bommeria* occur in North America, Mexico, and Central America. Only one species occurs in the United States.

B 1. *Bommeria hispida* (Mett. ex Kuhn)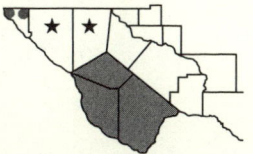
Underw., DANCING BOMMERIA, HAIRY BOM-
MER, COPPER FERN (*hispid* = hairy, bristly).
Fig. 12. [*Gymnopteris hispida* (Mett. ex Kuhn)
Underw.]. Infrequent in rocky, igneous soil on dry
mountain slopes and canyons in El Paso, Jeff Davis, Presidio, and Brewster Counties; also reported (Correll, 1955) from Hudspeth and Culberson Counties. 3,500–7,500 ft. NM, AZ. Mexico.

The only two members of the family Pteridaceae in the Trans-Pecos that have clearly pentagonal blades are *Bommeria hispida* and *Notholaena standleyi*. The undersides of the pinnae of dancing bommeria are decorated with scales and both coiled and straight hairs, while the pinnae undersides of Standley cloak fern have copious yellow or greenish farina but no hairs or scales.

5. *CHEILANTHES* Sw., LIP FERN

Members of the genus *Cheilanthes* range from 4 to 60 cm tall and usually grow on rock in dry habitats. The stems are rhizomatous, compact to long-creeping, with entire to toothed scales that are brown, black, or bicolored with dark central stripes and lighter margins. The leaf petioles are straw colored, brown or black; they are round, flat, or grooved adaxially and glabrous, scaly, or pubescent, with a single vascular bundle. The leaf blades are linear-lanceolate, ovate to elongate-pentagonal, pinnate-pinnatifid to

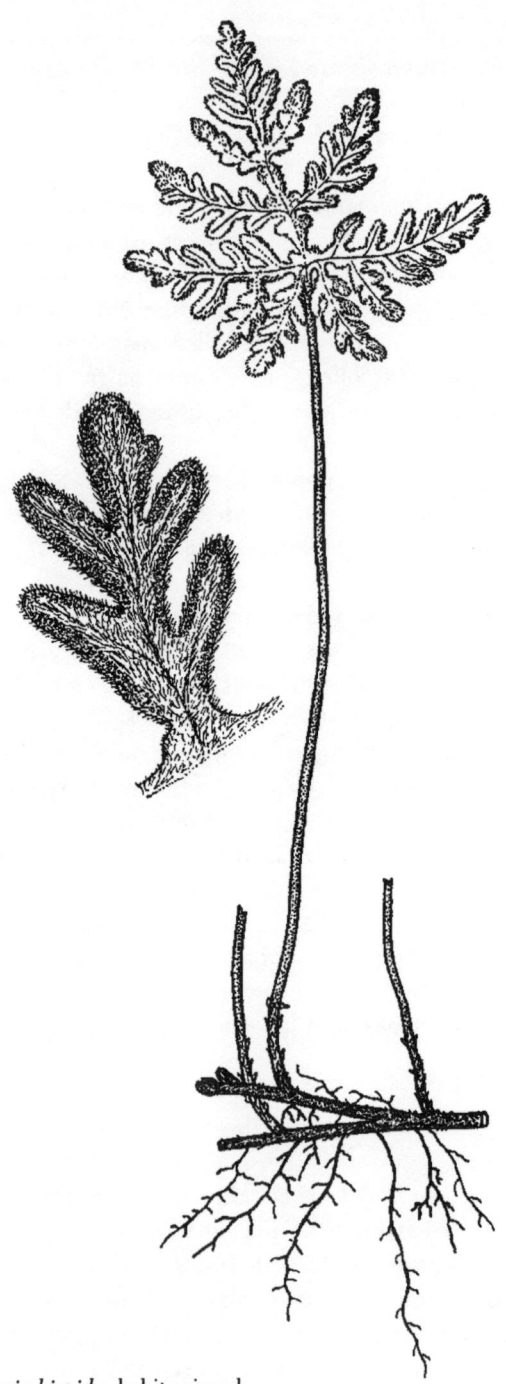

Fig. 12. *Bommeria hispida*, habit; pinnule.

4-pinnate. The abaxial surface may either be glabrous or have scales or hairs. Farina is never present. The adaxial (upper) surface is glabrous to variously pubescent. The pinnules are round to oblong, with the bases being rounded, truncate, or cuneate. The pinnules are generally less than 4 mm wide and are often free from the costae (pinnae rachis). The false indusia are marginal and narrow and are formed by reflexed segment margins. The sporangia occur at the vein tips near the margins.

The name *Cheilanthes* comes from the Greek *cheilos* for margin and *anthus* for flower, referring to the occurrence of the sporangia near the pinnule margins. This large, diverse genus has been variously treated by fern experts. The similar and difficult to distinguish genera *Notholaena* and *Pellaea* have been included in *Cheilanthes* by some authors. Other authors have segregated the taxa of *Notholaena*, *Pellaea*, and *Cheilanthes* in a variety of ways. Recent work has contributed to the clarification of the relationships among the classic genera. Two new genera, *Argyrochosma* and *Astrolepis*, have been recognized and *Notholaena* has been redefined. As a result of taxonomic reevaluation, pubescent, nonfarinose species formerly placed in *Notholaena* by some authors have been assigned to *Cheilanthes*. The genus *Cheilanthes*, as presented here, closely follows Windham and Rabe (1993). Most of the 150 species now assigned to the current concept of *Cheilanthes* are distributed in the Western Hemisphere. Fifteen of the 28 species that occur in the United States are found in the Trans-Pecos, and 16 species occur in Texas.

Key to the Species

1. Pinna rachis (costa) with scales on the underside; crozier (fiddlehead) tips hooked, not coiled . . . . (2).
1. Pinna rachis without scales on the underside; crozier tips coiled. . . . . (9).

2 (1). Pinna upper surface covered with stiff, often pustulate hairs (with tiny blisters at the base) . . . . 7. *C. horridula.*
2. Pinna upper surface glabrous or otherwise pubescent, lacking stiff, pustulate hairs. . . . . (3).

3 (2). Costal scales linear, inconspicuous, often appearing hairlike and blending with the tomentose pubescence
. . . . 11. *C. tomentosa.*
3. Costal scales lanceolate to ovate, conspicuous, easily differentiated from pubescence if hairs are present . . . . (4).

4 (3). Margins of the costal scales entire to denticulate (finely
toothed), not ciliate. . . . . (5).

    4. Margins of the costal scales ciliate, especially near base . (7).

5 (4). Pinnae glabrous on upper and lower surfaces; rhizome scales
concolored, brown . . . . 6. *C. fendleri.*

    5. Pinnae pubescent on upper and/or lower surfaces; rhizome
scales bicolored with dark center and lighter margins. . . (6).

6 (5). Costal scales ovate-lanceolate, densely imbricate, completely
obscuring the lower surface of the pinnae; pinnae sparsely to
moderately villous (shaggy) with coarse hairs . 12. *C. villosa.*

    6. Costal scales linear-lanceolate to lanceolate, moderately
imbricate, not obscuring the lower surface of the pinnae;
pinnae usually densely tomentose with fine, matted,
tangled hairs (some collections sparsely tomentose
above). . . . . 4. *C. eatonii.*

7 (4). Pinnae upper surface appearing densely tomentose;
pinnules less than 1 mm across; cilia of costal scales fine,
curly, tangled . . . . 10. *C. lindheimeri.*

    7. Pinnae upper surface appearing glabrous or sparsely
pubescent; some pinnules larger than 1 mm across; cilia
of costal scales coarse, not much tangled . . . . (8).

8 (7). Pinnae upper surface glabrous; costal scales ciliate toward the
base; rhizome scales concolored, brown, loosely appressed,
often deciduous with age. . . . . 13. *C. wootonii.*

    8. Pinnae upper surface often appearing pubescent (due to long,
curly cilia of costal scales); costal scales ciliate their entire
length; rhizome scales bicolored, dark brown, appressed and
persistent . . . . 15. *C. yavapensis.*

9 (1). Petioles distinctly grooved on adaxial side for most of their
length; pinnules glabrous on upper and lower surfaces
. . . . 14. *C. wrightii.*

    9. Petioles never grooved below the middle on the adaxial side;
pinnae glabrous, pubescent, or glandular . . . . (10).

10 (9). Rachis pubescence dimorphic, adaxially with dense, tortuous (twisted), appressed hairs, abaxially sparsely pubescent with long, often divergent (broadly spreading) hairs; pinnae sparsely hairy to glabrous . . . . (11).

10. Rachis pubescence monomorphic, hairs similar both ad- and abaxially; pinnae undersides conspicuously pubescent or glandular beneath . . . . (12).

11 (10). Blades broadly ovate to deltate in shape; rare plants mainly of Val Verde Co. . . . . 1. C. aemula.

11. Blades lanceolate to linear-oblong in shape; widely distributed . . . . 2. C. alabamensis.

12 (10). Pinnules nearly glabrous on upper surface, beadlike with false indusia forming a pouch with a constricted aperture on underside . . . . 9. C. lendigera.

12. Pinnules pubescent or glandular on upper surface; false indusia various, not forming a pouch as above . . . . . (13).

13 (12). Blades entirely pinnate-pinnatifid; pinnae articulate, stalk color not continuing into pinnule, underside usually densely golden tomentose, upper side hirsute with rather coarse, stiff hairs . . . . 3. C. bonariensis.

13. Blades 2–4-pinnate, at least at the base; pinnae not articulate, color of stalk continuing into pinnule underside . . . . . (14).

14 (13). Blades linear to broadly lanceolate, generally less than 5 cm wide; basal pinnae similar in size to the adjacent pair . . . . 5. C. feei.

14. Blades elongate-pentagonal, generally more than 5 cm wide; basal pinnae much larger than the adjacent pair . . . . 8. C. kaulfussii.

1. *Cheilanthes aemula* Maxon, TEXAS LIP FERN, RIVAL LIP FERN (*aemul* = emulating, rivaling). Fig. 13. In the Trans-Pecos known from only a few locations; Lake Amistad area, Seminole Canyon State Historic Park, and Devils River, Val Verde Co.; also listed in Big Bend Ranch State Park, Presidio and Brewster Counties. Known in the United States only from about 10 localities in Texas, most in the Edwards Plateau. 1,000–4,000 ft. TX. Mexico.

Cheilanthes aemula is similar to *C. alabamensis*, differing mainly in blade shape and spore number (64 for *aemula* and 32 for *alabamensis*). The blades of *C. aemula* are ovate to deltate, while those of *C. alabamensis* are lanceolate to narrow-oblong.

B,G 2. *Cheilanthes alabamensis* (Buckley) Kunze, ALABAMA LIP FERN (for the state of AL). Fig. 13. [*Pellaea alabamensis* (Buckley) Hook.]. Infrequent, usually on limestone, occasionally igneous substrates, in crevices, ledges, 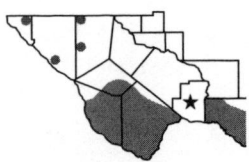 canyons, and slopes in El Paso, Hudspeth, Culberson, Jeff Davis, Presidio, Brewster, and Val Verde Counties; also reported (Correll, 1955) from Terrell Co. 1,000–6,000 ft. C TX, W to AZ, N to KS, E to VA, GA, not Gulf Coast. Mexico.

An unusual character that *C. alabamensis* and *C. aemula* have in common is the presence of abundant short, curly hairs on the adaxial side of the rachis. The abaxial side of the rachis in both species is nearly glabrous with a few long, spreading hairs.

B   3. *Cheilanthes bonariensis* (Willd.) Proctor, BONAIRE LIP FERN, GOLDEN LIP FERN (*bonaria* = in Buenos Aires). Fig. 14. [*Notholaena aurea* (Poir.) Desv.]. Infrequent to frequent on rocky bluffs and ledges in the Davis Mts., Jeff 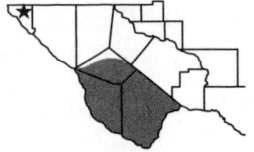 Davis Co.; Chinati Mts., Presidio Co.; Chisos and Glass Mts., Elephant Mt., and Sunny Glen, Brewster Co; also reported (Correll, 1955) from El Paso Co. 4,000–4,400 ft. NM, AZ. Mexico. West Indies. South America. Central America.

Cheilanthes bonariensis, with its pinnate-pinnatifid leaves, strongly resembles *Astrolepis sinuata* in habit. However, *C. bonariensis* lacks the abaxial scales of *A. sinuata*, having instead copious, usually golden, tomentose hairs on the underside of the pinnules. *Cheilanthes bonariensis* is the most abundant fern in west-central Mexico (Mickel, 1992) and is reportedly widely used medicinally in that country for stomach aches and coughs.

B,G 4. *Cheilanthes eatonii* Baker in Hook. & Baker, EATON LIP FERN (for A. Eaton, noted American botanist). Fig. 15. [*C. castanea* Maxon; *C. eatonii* forma *castanea* (Maxon) Correll]. Infrequent to frequent on igneous, limestone, or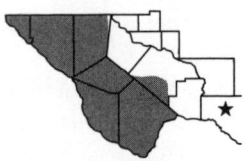

Fig. 13. *Cheilanthes alabamensis*, habit; pinnule. *C. aemula*, frond; pinnule.

sandstone substrates on rocky slopes, ledges, and at the bases of boulders in mountainous areas of El Paso, Hudspeth, Culberson, Jeff Davis, Presidio, Brewster, and Pecos Counties; also reported (Correll, 1955) from Val Verde Co. (1,200) 3,000–7,500 ft. C and N TX, W to AZ, N to CO, E to AR, VA. Mexico. Central America.

Cheilanthes eatonii is one of the most common ferns in mountainous areas of the Trans-Pecos. The pubescence on the upper side of the pinnae of C. *eatonii* varies from the typically dense villous to virtually glabrous.

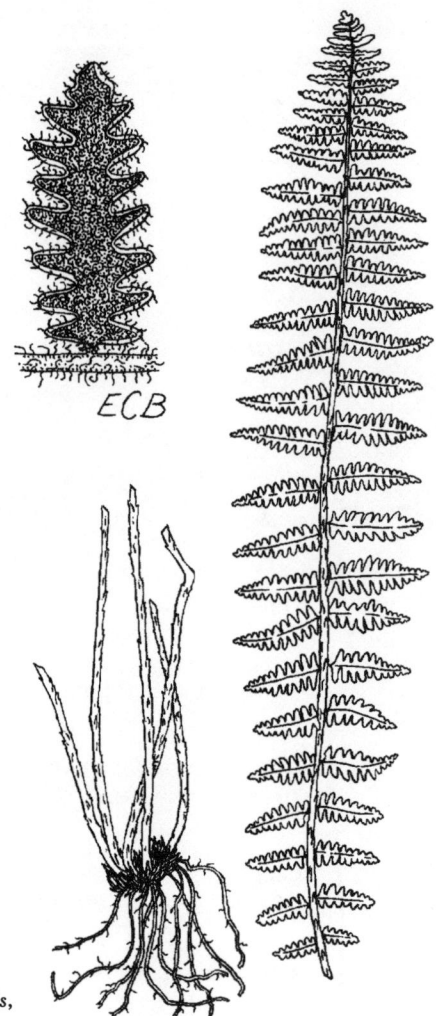

Fig. 14. *Cheilanthes bonariensis*, habit; pinnule.

The glabrous collections have been designated by some fern taxonomists as *C. castanea* or *C. eatonii* forma *castanea*. Until further study clarifies the relationships within the *C. eatonii* complex, these morphologically varied specimens are treated here as one taxon.

B,G 5. *Cheilanthes feei* T. Moore, SLENDER LIP FERN (for A. Fee, noted French botanist). Fig. 15. Infrequent to frequent on ledges, cliffs, crevices, and crags, usually on limestone, occasionally on igneous substrates, in the mountainous

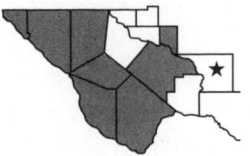

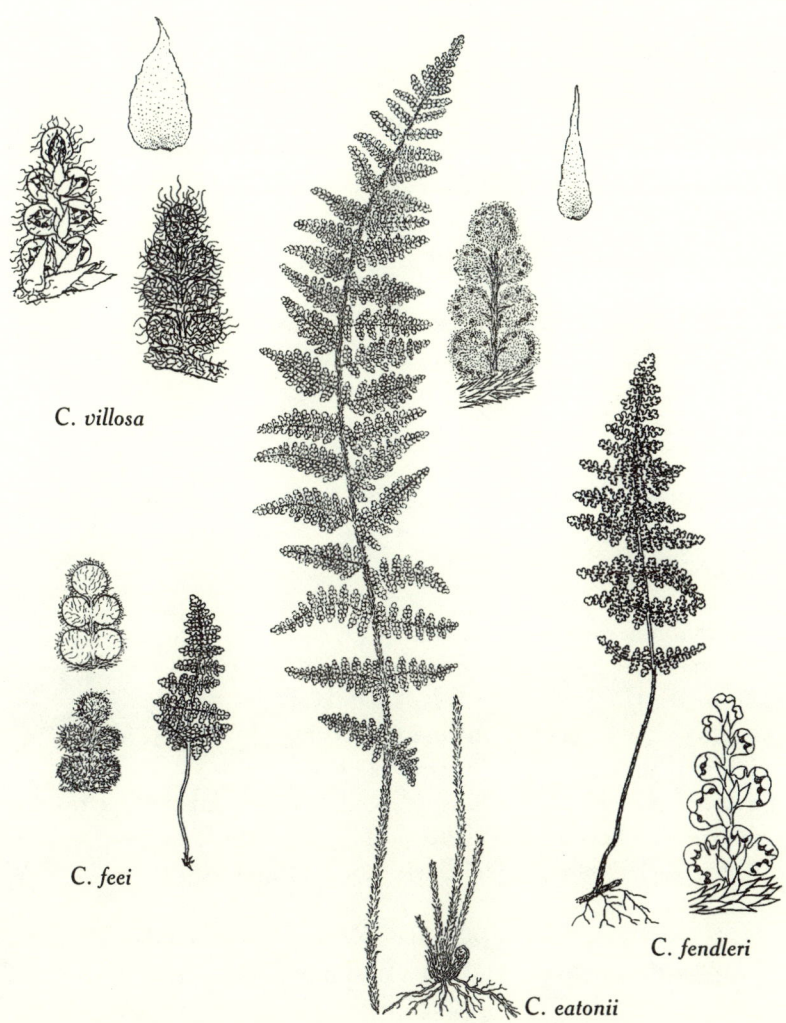

Fig. 15. *Cheilanthes feei*, frond; pinnule, lower surface, upper surface.
C. *villosa*, pinnule, lower surface, upper surface; costal scale.
C. *eatonii*, habit; pinnule; costal scale. C. *fendleri*, habit; pinnule.

areas of El Paso, Hudspeth, Culberson, Jeff Davis, Presidio, Brewster, Pecos, and Crane Counties; reported but unconfirmed in Big Bend National Park, Brewster Co. (National Park Service, 1995); also reported (Correll, 1955) from Crockett Co. 2,800–8,000 ft. N, C, and W TX, N to MN, SD, MT; W to CA, OR, WA; E to IL, WI, AR. Alberta, British Columbia, Canada. Mexico.

Cheilanthes feei is most abundant in the plains states, where it is often found in dry habitats. At first glance, C. *feei* may resemble small plants of C. *tomentosa*. These species are most easily distinguished by the stipe pubescence, which in C. *feei* consists of crisp, multicellular, spreading hairs, with no linear scales, while C. *tomentosa* has loosely appressed, single-celled hairs mixed with linear scales.

6. *Cheilanthes fendleri* Hook., FENDLER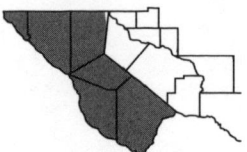
LIP FERN (for A. Fendler, noted plant collector).
Fig. 15. Infrequent on slopes, ledges, and at the
base of boulders, on various but usually igneous
substrates in El Paso, Hudspeth, Culberson, Jeff
Davis, Brewster, and Presidio Counties. 4,000–8,000 ft. NM, AZ, CO. Mexico.

The combination of at least slightly imbricate, ovate, nonciliate costal scales and complete lack of other pubescence distinguishes C. *fendleri* from the other species of *Cheilanthes*.

B   7. *Cheilanthes horridula* Maxon, PRICKLY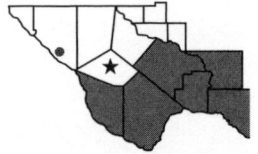
LIP FERN, ROUGH LIP FERN (*horrid* = rough,
prickly). Fig. 16. [*Pellaea aspera* (Hook.) Baker].
Infrequent to frequent, usually on limestone, occasionally on igneous substrates, in crevices or soil,
on ledges, bluffs, and at the base of boulders in Hudspeth, Presidio, Brewster, Pecos, Terrell, Crockett, and Val Verde Counties; also reported (Correll, 1955) from Jeff Davis Co. 1,000–4,500 ft. S-C and W TX, OK.

Prickly lip fern is the only species of *Cheilanthes* in which the pinnae are covered with stiff, rough, prickly, often pustulate hairs.

B   8. *Cheilanthes kaulfussii* Kunze, GLAND-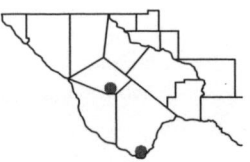
ULAR LIP FERN, KAULFUSS LIP FERN (for G.
Kaulfuss, noted German fern botanist). Fig. 17.
Infrequent and rare in crevices, on ledges, slopes,
and among boulders in the Chisos Mts., Brewster
Co., and an isolated occurrence near Mitre Peak in Jeff Davis Co., 1938. 4,500–7,500 ft. Edwards Plateau, TX. Mexico. Central America. South America.

Cheilanthes kaulfussii is covered with conspicuous glandular, capitate hairs that exude a sticky light brown oil when fresh. *Cheilanthes kaulfussii* resembles C. *leucopoda*, which occurs to the east of the Trans-Pecos on the Edwards Plateau. *Cheilanthes kaulfussii* has a dark stipe, whereas in C.

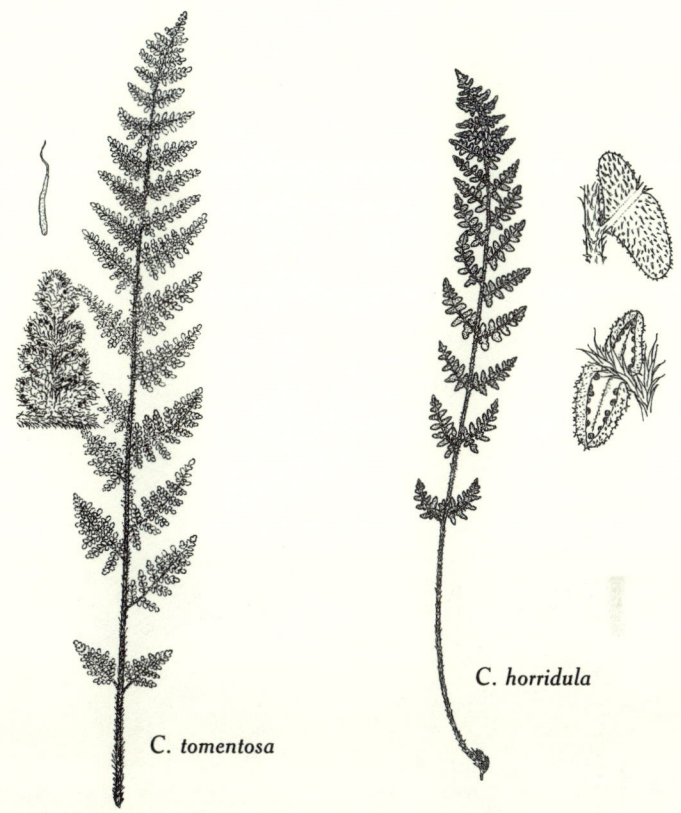

Fig. 16. *Cheilanthes tomentosa*, frond; pinnule; costal scale. *C. horridula*, frond; pinnule, upper surface, lower surface.

leucopoda the stipe is straw colored and the pubescence is of long hairs that are not capitate or glandular.

B   9. *Cheilanthes lendigera* (Cav.) Sw., BEADED LIP FERN, (*lend* = nit + *ger* = to bear or carry). Fig. 17. Very rare on rocky slopes and ledges on igneous substrates in the Chisos Mts., Brewster Co., 4,000–7,200 ft. AZ. Mexico. Central America. South America.

The underside of the pinnules of beaded lip fern have distinctive, pouchlike margins that are formed by the broad false indusia causing the tiny segments to appear beadlike.

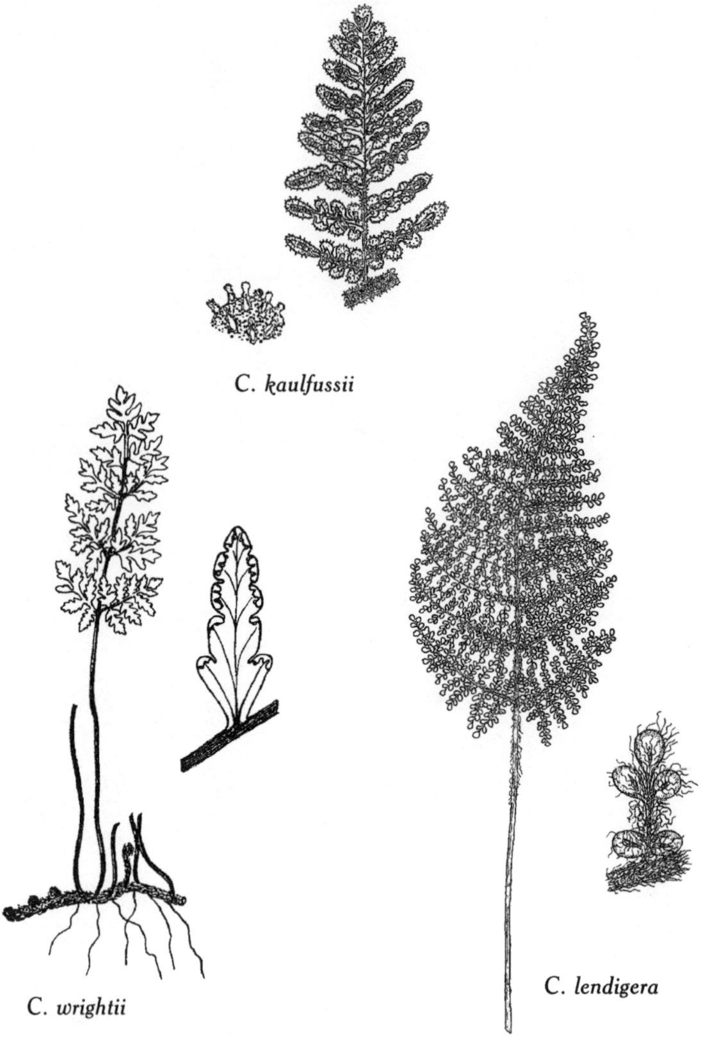

C. kaulfussii

C. wrightii

C. lendigera

Fig. 17. *Cheilanthes wrightii*, habit; pinnule. C. *kaulfussii*, pinna; pinnule. C. *lendigera*, frond; pinnule.

B   10. *Cheilanthes lindheimeri* Hook., Fairy Swords, Whitefoot Lip Fern (for F. Lindheimer, noted Texas plant collector). Fig. 18. Infrequent to frequent on ledges, cliffs, boulders, and in canyons on igneous and sandstone

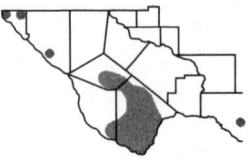

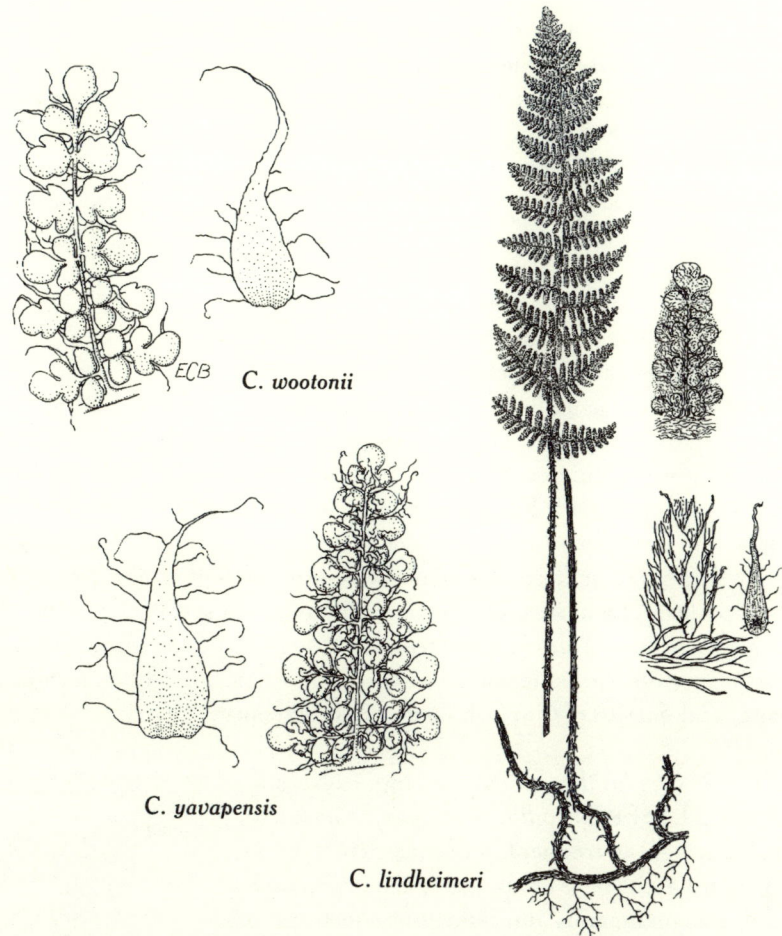

Fig. 18. *Cheilanthes yavapensis*, pinnule; costal scale. *C. wootonii*, pinnule; costal scale. *C. lindheimeri*, habit; pinnule, upper surface, lower surface; costal scale.

substrates in El Paso, Hudspeth, Jeff Davis, Presidio, Brewster, and Val Verde Counties, and in novaculite S of Marathon, Brewster Co. (1,100) 3,400–5,600 ft. Edwards Plateau and C TX, AZ, NM. Mexico.

Cheilanthes lindheimeri, *C. wootonii*, and *C. yavapensis* are quite similar in appearance and are best distinguished by carefully noting the subtle key characters. The glabrous or sparsely pubescent upper pinnule surfaces of *C. lindheimeri* may have a tomentose appearance because the long curly cilia of the abaxial scales appear to be tomentose hairs on the upper surface. The cilia of the scales of *C. yavapensis* are less fine and tangled but may nevertheless appear to be pubescence on the upper side of the actually glabrous pinnules. The scales of C. *wootonii* are ciliate only toward the base, and consequently the upper surface of the pinnules appears to be glabrous. The ultimate segments of all three species are generally round to oblong, with the segments of C. *lindheimeri* being less than 1 mm in diameter, those of C. *wootonii* being 1–3 mm, and those of C. *yavapensis* being 1–2 mm. It may not always be possible to identify these species with complete certainty.

B 11. *Cheilanthes tomentosa* Link, WOOLLY LIP FERN (*toment* = dense matted wool). Fig. 16. Infrequent on ledges, cliffs, slopes, and at the base of boulders usually on igneous substrates in mountainous areas of Jeff Davis, Brewster, Pecos, 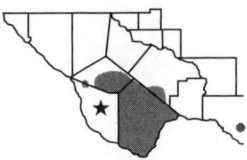 and Val Verde Counties; also reported from Presidio Co. (Correll, 1955). (1,100) 4,000–6,000 ft. Throughout much of Texas, except for the Panhandle and Gulf Coast areas. W to AZ, N to KS, E to VA, NC. Mexico.

Cheilanthes tomentosa is similar to C. *eatonii* except for the difference in costal scales. The narrow and inconspicuous costal scales of C. *tomentosa* often appear to intergrade into the tomentose hairs on the underside of the pinnules, while the scales of C. *eatonii* are broader, usually lanceolate in shape, and easy to distinguish from the pubescence.

B,G 12. *Cheilanthes villosa* Davenp. ex Maxon, VILLOUS LIP FERN (*villos* = shaggy, hairy). Fig. 15. Infrequent to frequent in crevices, cliffs, rocky slopes and at the base of boulders, usually on limestone, also on igneous and sandstone substrates in 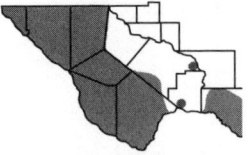 mountainous areas of El Paso, Hudspeth, Culberson, Jeff Davis, Presidio, Brewster, Pecos, Terrell, and Val Verde Counties. 2,100–6,000 ft. NM, AZ. Mexico.

The costal scales of C. *villosa* are broad and imbricate and virtually cover the bottom of the pinnules, often obscuring the undersurface completely. When the pinnule is viewed from the top, the tips of the scales are usually visible extending beyond the pinnule edge.

13. *Cheilanthes wootonii* Maxon, WOOTON LIP FERN (for E. O. Wooton, noted New Mexico botanist). Fig. 18. Rare on rocky slopes and ledges in the Franklin and Hueco Mts., El Paso Co.; 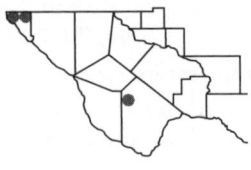 and Sunny Glen, Brewster Co. 4,000–5,000 ft. TX, NM, AZ, CO, UT, NV, CA, OK. N Mexico.

Cheilanthes *wootonii* is closely related to C. *lindheimeri* and C. *yavapensis*, which are similar in appearance. Distinguishing characters of these three species are discussed in the C. *lindheimeri* species account.

B   14. *Cheilanthes wrightii* Hook., WRIGHT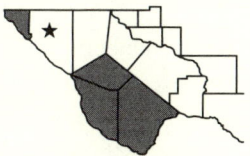
LIP FERN (for C. Wright, noted plant collector).
Fig. 17. Infrequent on igneous or novaculite
substrates in mountainous areas of El Paso, Jeff
Davis, Presidio, and Brewster Counties; also
reported (Correll, 1955) from Hudspeth Co. 3,500– 6,500 ft. NM, AZ.
Mexico.

Cheilanthes *wrightii* is generally a small fern that is best distinguished from other *Cheilanthes* species by the grooved and glabrous (to sparsely hairy and occasionally scaly) stipe and the glabrous pinnae, which lack costal scales. Similar members of the genus *Pellaea* do not have a grooved rachis. *Cheilanthes aemula* and C. *alabamensis*, which may have glabrous pinnules, have distinctive dimorphic hairs on an ungrooved stipe.

15. *Cheilanthes yavapensis* T. Reeves ex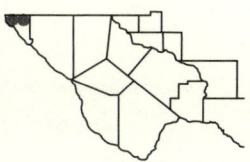
Windham, YAVAPAI LIP FERN, GRACEFUL LIP
FERN (for the type locality in Yavapai Co., AZ).
Fig. 18. Rare in the Franklin and Hueco Mts., El
Paso Co. 3,500–4,500 ft. NM, AZ.

Because the morphological characters that separate C. *yavapensis* from C. *wootonii* are small and subtle, it may not be posssible to identify all collections with certainty. Windham and Rabe (1993) believe that the similarities of these two species are attributable to hybrid convergence rather than to common ancestry.

6. *NOTHOLAENA* R. Brown, CLOAK FERN

Plants of the genus *Notholaena*, which can grow to be 35 cm high, are usually found in xeric habitats growing on rock. The stems are rhizomatous, short-creeping to compact, with scales that are toothed to entire, concolored black or bicolored with a dark central stripe and lighter margins. The leaf petioles are brown or black, with or without an adaxial longitudinal groove. Scales, farinose glands, or hairs are usually present. The vascular bundle is single. The leaf blades are linear-lanceolate to pentagonal, pinnate-pinnatifid to 4-pinnate, with white, yellowish, or greenish farina present on the abaxial surface. In *N. aschenborniana* the farina is mostly obscured by scales. The adaxial surface of the blade is glabrous, variously pubescent, or glandular. Pinnules are elliptic to oblong-ovate, generally less than 4 mm wide, and often adnate to the costae. The false indusia are marginal, narrow, and formed by reflexed segment margins. The sori occur at the vein tips near the margins.

The name *Notholaena* is derived from the Greek *notho* for spurious or false and *chlaena* for cloak, referring to the lack of a true indusium. The current treatment follows Windham (1993a), who assigned pubescent, nonfarinose former *Notholaena* species to *Cheilanthes;* scaly, nonfarinose species to *Astrolepis;* and glabrous, farinose species, which are closely related to *Pellaea,* to *Argyrochosma.* The 25 species currently recognized for *Notholaena* are distributed in North America, Central America, South America, Mexico, and the West Indies. Eight of the 10 species that occur in the United States are found in Texas and in the Trans-Pecos.

Key to the Species

1. Farina present on the undersides of pinnae but concealed by scales and hairs . . . . 2. *N. aschenborniana.*
1. Farina apparent on undersides of pinnae . . . . (2).

2 (1). Scales absent on the blades. . . . . (3).
   2. Scales present on the blades, lanceolate to needlelike . . (6).

3 (2). Upper surfaces of pinnae densely glandular; blades narrowly deltate; petioles grooved adaxially. . . . . 5. *N. greggii.*
   3. Upper surfaces of pinnae glabrous to sparsely glandular; blades ovate to pentagonal; petioles rounded adaxially . . (4).

4 (3). Blades pentagonal, about as broad as they are long
   . . . . 8. *N. standleyi.*
   4. Blades ovate, longer than they are broad . . . . (5).

5 (4). Farina on undersides of pinnae stark white; blades at most bipinnate . . . . 3. *N. copelandii.*
   5. Farina on undersides of pinnae yellowish white; blades tripinnate (3 times dissected) . . . . 7. *N. neglecta.*

6 (2). Blade scales shiny, resembling needlelike hairs 6. *N. nealleyi.*
   6. Blade scales dull, flat, lanceolate to linear-lanceolate. . . (7).

7 (6). Upper pinnae surfaces villous with long, delicate, white hairs,
sparsely glandular; blade scales ciliate-dentate . 1. *N. aliena.*
7. Upper pinnae surfaces lacking white hairs, surface mostly
glabrous with scattered glandular pubescence; blade scales
generally entire . . . . 4. *N. grayi.*

B   1. *Notholaena aliena* Maxon, MEXICAN
CLOAK FERN, FOREIGN CLOAK FERN (*alien* =
foreign, possibly referring to its occurring mostly
in Mexico). Fig. 19. [*Cheilanthes aliena* (Maxon)
Mickel]. A very rare species reported to occur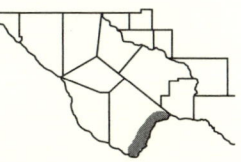
(Correll, 1955; *Flora of North America,* 1993) on rocky slopes and cliffs
usually on volcanic substrates in the Chisos Mts., Black Gap Wildlife
Management Area, and the lower canyons of the Rio Grande, Brewster
Co. 1,600–3,000 ft. Mexico.

Notholaena aliena with its distinctly white villous upper pinnules is easily distinguished from the closely related *N. grayi,* which has pinnules that
are usually glandular and totally lacking in hairs.

2. *Notholaena aschenborniana* Klotzsch,
ASCHENBORN CLOAK FERN, SCALED CLOAK
FERN. Fig. 19. [*Cheilanthes aschenborniana*
(Klotzsch) Mett.]. Infrequent on limestone sub-
strates in the Glass and Del Norte Mts., Brewster
Co.; Pecos Co.; reported (Correll, 1955) from Val Verde Co. 2,200–
5,000 ft. AZ. Mexico.

Although the blades of *N. aschenborniana* have a thick coating of farina
on the lower surface, the farina is not readily apparent as it is usually completely concealed by overlapping scales and hairs.

B   3. *Notholaena copelandii* C. C. Hall,
COPELAND CLOAK FERN (for E. B. Copeland,
noted fern botanist). Fig. 20. [*Cheilanthes*
candida var. *copelandii* (C. C. Hall) Mickel;
Notholaena candida var. *copelandii* (C. C. Hall)
R. M. Tryon]. Frequent to infrequent in canyons, on ledges, and on slopes
in limestone substrates in Del Norte Mts., Brewster Co.; Terrell Co.; canyons along the Rio Grande in Brewster and Val Verde Counties; reported
but unconfirmed in Big Bend National Park, Brewster Co. (National
Park Service, 1995); also reported (Correll, 1955) from Pecos Co.
1,000–3,700 ft. Edwards Plateau, TX. Mexico.

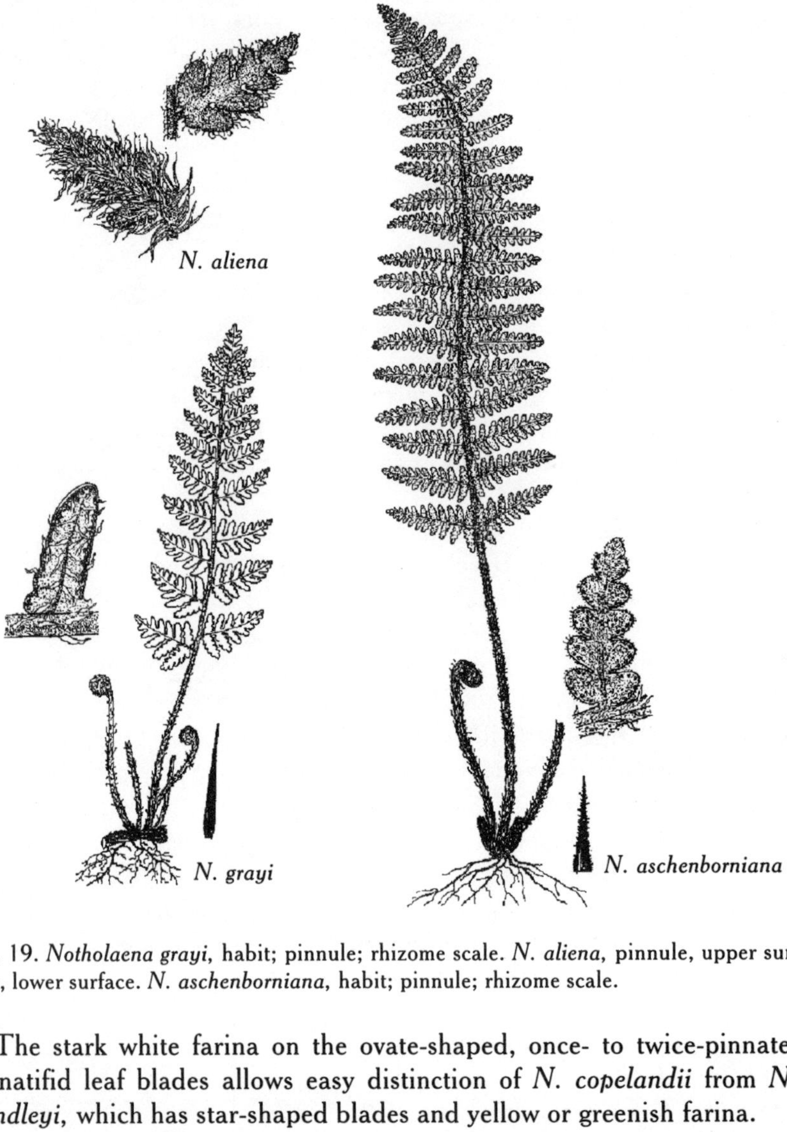

N. aliena

N. grayi

N. aschenborniana

Fig. 19. *Notholaena grayi*, habit; pinnule; rhizome scale. *N. aliena*, pinnule, upper surface, lower surface. *N. aschenborniana*, habit; pinnule; rhizome scale.

The stark white farina on the ovate-shaped, once- to twice-pinnate-pinnatifid leaf blades allows easy distinction of *N. copelandii* from *N. standleyi*, which has star-shaped blades and yellow or greenish farina.

B   4. *Notholaena grayi* Davenp. ssp. *grayi*, GRAY CLOAK FERN (for A. Gray, noted botanist). Fig. 19. [*Cheilanthes grayi* (Davenp.) Domin.]. Rare to frequent in Sunny Glen, near Alpine, and in the Basin of the Chisos Mts., Brewster Co.; Big Bend Ranch State Park, Brewster and Presidio Counties (Worthington, 1995) on granite or limestone substrates; also

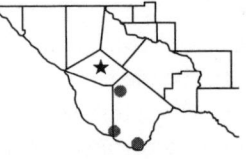

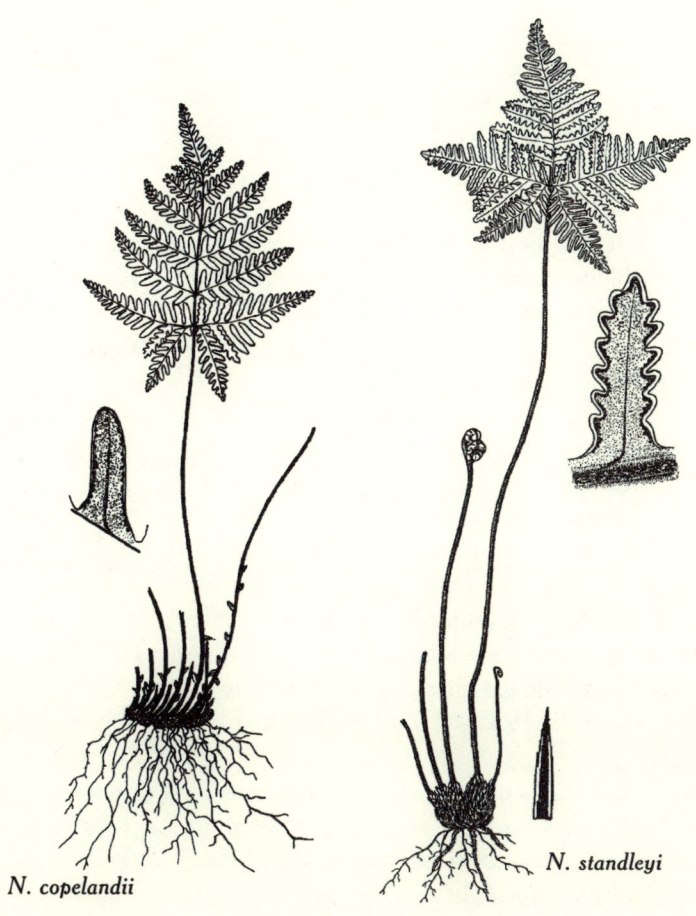

Fig. 20. *Notholaena copelandii,* habit; pinnule. *N. standleyi,* habit; pinnule; rhizome scale.

reported from Jeff Davis Co. (Correll, 1955). 4,000–6,000 ft. C TX, NM, AZ. Mexico.

Notholaena grayi is similar to *N. aliena,* differing in the presence of long, multicellular whitish hairs on the upper pinnae surfaces in *N. aliena,* while the upper pinnae surfaces of *N. grayi* are glabrous or slightly glandular. Also the margins of the scales on the lower surfaces of the pinnae of *N. grayi* are generally entire, while those of *N. aliena* are ciliate-toothed, a character that is difficult to see even under magnification.

B   5. *Notholaena greggii* (Mett. ex Kuhn)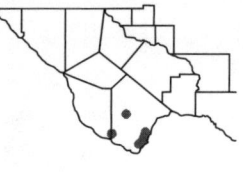
Maxon, GREGG CLOAK FERN (for J. Gregg,
noted botanical collector). Fig. 21. [*Cheilanthes
greggii* (Mett. ex Kuhn) Mickel]. Rare on limestone bluffs, slopes, and canyons in the Big Bend
Region of S Brewster Co. 2,000–3,800 ft. Mexico.

Gregg cloak fern is best distinguished from the other species of *Notholaena* that lack blade scales by its densely glandular upper pinnae surfaces
and adaxially grooved petioles. Although the farina on the underside of the
pinnules may be sparse, a careful examination will show that at least some
farina is present in *N. greggii*.

B   6. *Notholaena nealleyi* Seaton ex J. M.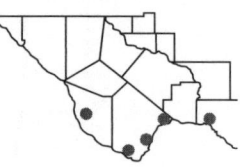
Coult., NEALLEY CLOAK FERN (for G. C.
Nealley, noted plant collector in the U.S. South-
west and Texas). Fig. 21. [*Cheilanthes nealleyi*
(Seaton ex J. M. Coult.) Domin; *Notholaena
schaffneri* var. *nealleyi* (Seaton ex J. M. Coult.) Weath.]. Rare on lime-
stone cliffs, ledges, crevices, and talus slopes near the Rio Grande in
Brewster and Val Verde Counties; the type locality is in the Chinati Mts.,
Presidio Co. (Correll, 1955). 1,200–4,000 ft. Edwards Plateau and S
TX. Mexico.

Nealley cloak fern has unique, shiny blade scales that resemble needlelike hairs. *Notholaena nealleyi* is closely related to *N. schaffneri*, but
Windham (1993a) recognized *N. nealleyi* as distinct on the basis of ge-
netic studies, morphological differences, and geographical isolation.

    7. *Notholaena neglecta* Maxon, MAXON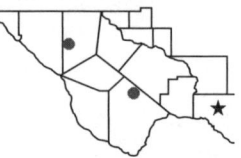
CLOAK FERN (*neg* = refuse, deny + *lego* =
choose, gather). Fig. 21. [*Cheilanthes neglecta*
(Maxon) Mickel]. Rare on limestone slopes in the
Sierra Diablo Mts., Culberson Co.; Glass Mts.,
Brewster Co.; also reported (Correll, 1955) from Val Verde Co. (2,000)
4,500–5,000 ft. Edwards Plateau, TX, AZ. Mexico.

The blades of *N. neglecta* usually are more pentagonal in shape and
more finely dissected than are those of the similar *N. copelandii*. The farina
of *N. copelandii* is distinctly stark white when compared to the yellowish farina of *N. neglecta*.

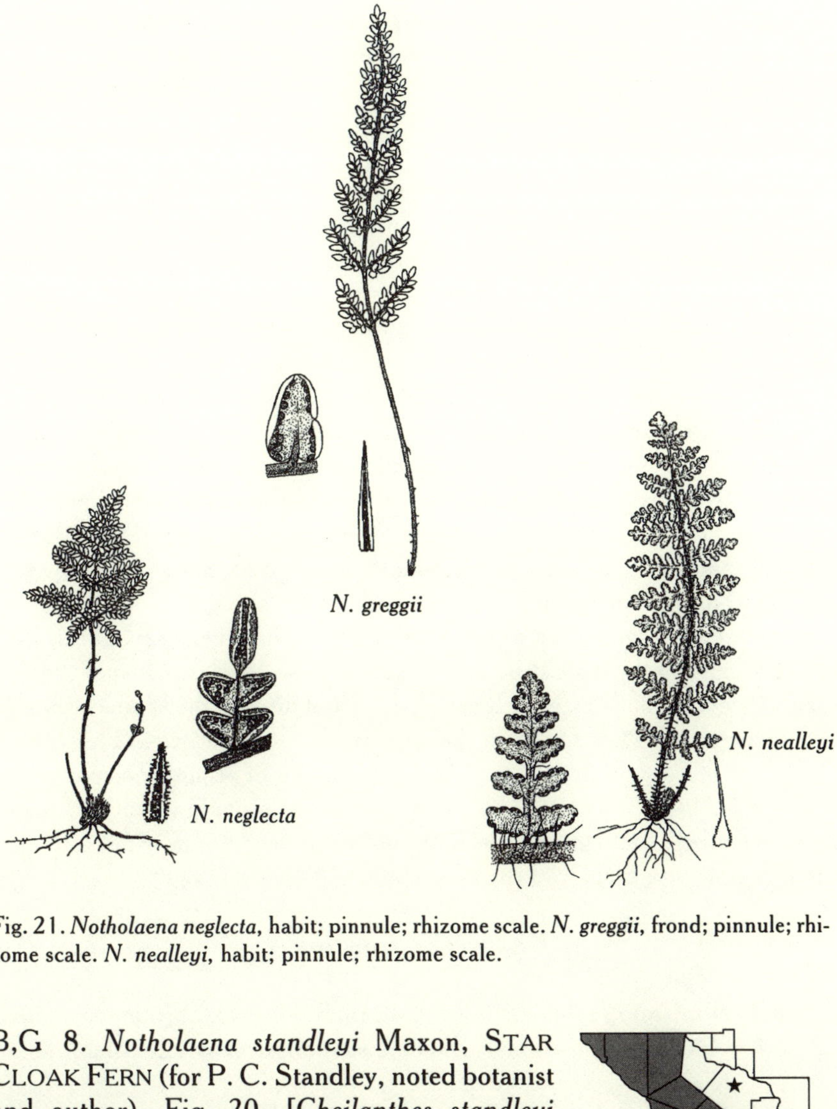

Fig. 21. *Notholaena neglecta*, habit; pinnule; rhizome scale. *N. greggii*, frond; pinnule; rhizome scale. *N. nealleyi*, habit; pinnule; rhizome scale.

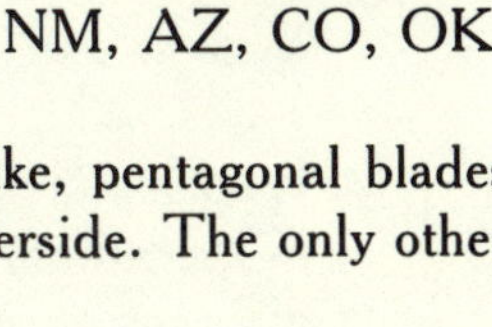

B,G 8. *Notholaena standleyi* Maxon, STAR CLOAK FERN (for P. C. Standley, noted botanist and author). Fig. 20. [*Cheilanthes standleyi* (Maxon) Mickel]. Infrequent to abundant in crevices, crannies, among boulders in canyons, on 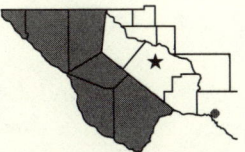 slopes, ledges, and bluffs in igneous and limestone substrates, throughout much of the Trans-Pecos. 1,600–6,800 ft. TX, NM, AZ, CO, OK. Mexico.

Notholaena standleyi is recognized by its starlike, pentagonal blades with copious yellow or greenish farina on the underside. The only other

Trans-Pecos fern that has leaves of a similar shape is *Bommeria hispida*, which has hairy rather than farinaceous leaf blade undersides.

7. *PELLAEA* Link, CLIFF BRAKE

Members of the genus *Pellaea* may grow to 1 m high and are usually found in xeric habitats on rock. The stems are rhizomatous, short and stout to long-creeping, with scales that are toothed to entire, concolored brown or tan, or bicolored with a dark central stripe and lighter margins. The leaf petioles, which have a single vascular bundle, are black, brown, yellow, or tan and often have sparse basal scales. The leaf blades are linear to ovate-deltate, 1–4-pinnate toward the base, glabrous or sparsely pubescent, and often bluish gray or green. Some species (not including those from the Trans-Pecos) have farina on the underside of the pinnules. The pinnules are linear, lanceolate to elliptic, generally stalked, and free from the costae. The false indusia are marginal, narrow, and formed by reflexed segment margins. The sori occur along the ends of the veins near the segment margins and often form a continuous marginal band.

The name *Pellaea* is from the Greek *pellos*, meaning dusky, and is believed to refer to the dark stipe and rachis of some species. Most of the 40 species occur in the Western Hemisphere, 15 of them in the United States, and seven species are found in Texas and the Trans-Pecos. The genus *Pellaea* is closely related to the glabrous species of *Argyrochosma*, from which it can be differentiated by a combination of pinnule width and scale color. Members of *Pellaea* have pinnules more than 4 mm wide, often in combination with bicolored rhizome scales.

Key to the Species

1. Stipe and rachis yellow or tan; pinnules oval, elliptic, or oblong, round-tipped; rhizome scales narrowly lanceolate to ovate. . . . . (2).
1. Stipe and rachis brown to black; pinnules elongate, usually pointed; rhizome scales narrowly linear. . . . . (4).

2 (1). Rachis and costae zigzag; pinnae often pointing downward; rhizome scales bicolored; leaf blades usually glabrous . . . . 4. *P. ovata.*
2. Rachis and costae straight; pinnae perpendicular to the rachis or ascending; rhizome scales con- or bicolored; leaf blades glabrous or pubescent . . . . (3).

3 (2). Pinnules round to deltate, most with distinctly cordate bases; rhizome scales concolored; scales on lower stipe ovate-lanceolate; rachis and costae essentially glabrous
. . . . 2. *P. cordifolia.*

   3. Pinnules elliptic to ovate, base not distinctly cordate; rhizome scales bicolored; scales on lower stipe narrow lanceolate; rachis and costae usually pubescent . . . . . 3. *P. intermedia.*

4 (1). Rhizome scales uniformly red-brown or tan, not bicolored; stipe and rachis moderately pubescent with appressed crisp hairs . . . . 1. *P. atropurpurea.*

   4. Rhizome scales bicolored, with dark central region and lighter brown margin; stipe and rachis glabrous or variously pubescent. . . . . (5).

5 (4). Leaf blades deeply pinnate-pinnatifid toward the base, basal pinnae ternately lobed (palmate with 3 leaflets); petioles generally dark purple to black. . . . . 5. *P. ternifolia.*

   5. Leaf blades 2–3-pinnate toward the base; petioles generally brown or reddish brown . . . . (6).

6 (5). Total length of pinna costae much longer than largest individual pinnule; largest pinnae usually divided into 9–11 segments; leaf blade generally more than 4.5 cm wide, ovate-deltate to lanceolate in shape . . . . . . 6. *P. truncata.*

   6. Total length of pinna costae shorter than, or equal to, largest individual pinnule; largest pinnae usually divided into 3–5 segments; leaf blade generally less than 4.5 cm wide, linear-oblong in shape. . . . . 7. *P. wrightiana.*

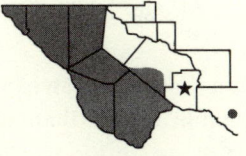

B,G   1. *Pellaea atropurpurea* (L.) Link, PURPLE CLIFF BRAKE (*atr* = black + *purpur* = purple, for the purple-black stipe and rachis). Fig. 22. [*Pteris atropurpurea* (L.) Link]. Abundant to infrequent in canyons and open rocky woods, on mountain slopes, cliff faces, ledges, and among rocks on limestone and igneous substrates throughout much of the Trans-Pecos. (1,200) 3,500–7,000 ft. Most of TX, except Gulf Coast, most of the E, C, and SW U.S. Mexico. Central America.

   The slightly dimorphic blades of *P. atropurpurea* are bipinnate toward the base, and the linear-oblong pinnules of the fertile blades are narrower

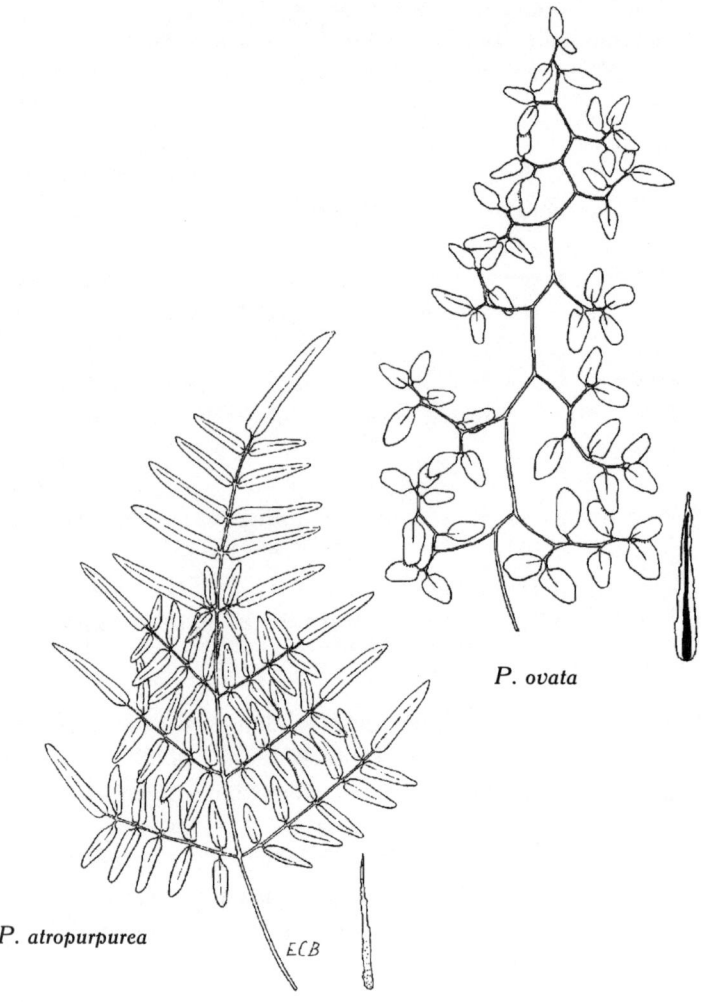

P. ovata

P. atropurpurea   ECB

Fig. 22. *Pellaea atropurpurea*, frond; rhizome scale. *P. ovata*, frond; rhizome scale.

than those of the sterile blades. Often the terminal pinnule of a pinna is noticeably longer than the lateral pinnules. The narrowly linear rhizome scales are uniformly rust-brown or tan and lack a dark central stripe, and the rachis and costae are pubescent with appressed, crisp hairs. Purple cliff brake is possibly the most widely distributed fern in Texas, growing in many areas where rock outcrops occur. *Pellaea atropurpurea* is quite hardy and would make an attractive addition to a rock garden.

B   2. *Pellaea cordifolia* (Sessé & Moc.) A. R.
Sm., HEARTLEAF CLIFF BRAKE (*cordi* = heart
+ *folia* = leaf, for the shape of the pinnules). Fig.
23. [*P. cardiomorpha* Weath.; *P. sagittata* var.
cordata (Cav.) A. F. Tryon]. Rare in canyons, on
cliffs, and on mountain slopes in the Davis Mts., Jeff Davis Co., and
Chisos Mts., Brewster Co.; also reported (Correll, 1955) from Presidio
Co. 6,000–7,800 ft. Mexico.

The ovate-deltate blades of *P. cordifolia* are bipinnate toward the base.
The heart-shaped pinnules with deeply cordate bases and the straight
rachis and uniformly colored rhizome scales distinguish *P. cordifolia* from
P. ovata. Heartleaf cliff brake, although common in Mexico, enters the
United States only at a few locations in the Trans-Pecos.

B,G 3. *Pellaea intermedia* Mett. ex Kuhn,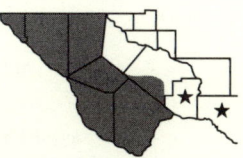
CREEPING CLIFF BRAKE, INTERMEDIATE CLIFF
BRAKE (*inter* = between + *medi* = middle). Fig.
23. Frequent to infrequent in canyons, on slopes,
and among boulders on igneous and limestone
substrates in El Paso, Hudspeth, Culberson, Jeff Davis, Presidio, Brewster, and Pecos Counties; reported from Terrell and Val Verde Counties
(Correll, 1955). (2,200) 3,100–7,000 ft. NM, AZ. Mexico.

The blades of creeping cliff brake are bipinnate with elliptic, oval, or oblong pinnules, which have truncate or broadly rounded bases and rounded
tips. The narrowly lanceolate rhizome scales have a black, thick center and
thinner, brown margins. The blade rachis and costae are pubescent with
short hairs that may disappear with age.

B   4. *Pellaea ovata* (Desv.) Weath., ZIGZAG
CLIFF BRAKE, OVATE LEAF CLIFF BRAKE
(*ovate* = eggshaped, for the shape of the
pinnules). Fig. 22. Frequent to infrequent among
boulders, on ledges and rocky slopes, and in can-
yons on limestone and other substrates in the Chisos Mts. and vicinity,
Brewster Co.; S of Sanderson, Terrell Co.; and along the Pecos River and
Devils River, Val Verde Co. 1,000–4,500 ft. C and S half of TX. Mexico. West Indies. Central America. South America.

The blades of *Pellaea ovata* are usually 3-pinnate toward the base and
the pinnules are ovate in shape. The zigzag rachis and costae of *P. ovata*
are distinctive and easily separate this taxon from the other Trans-Pecos

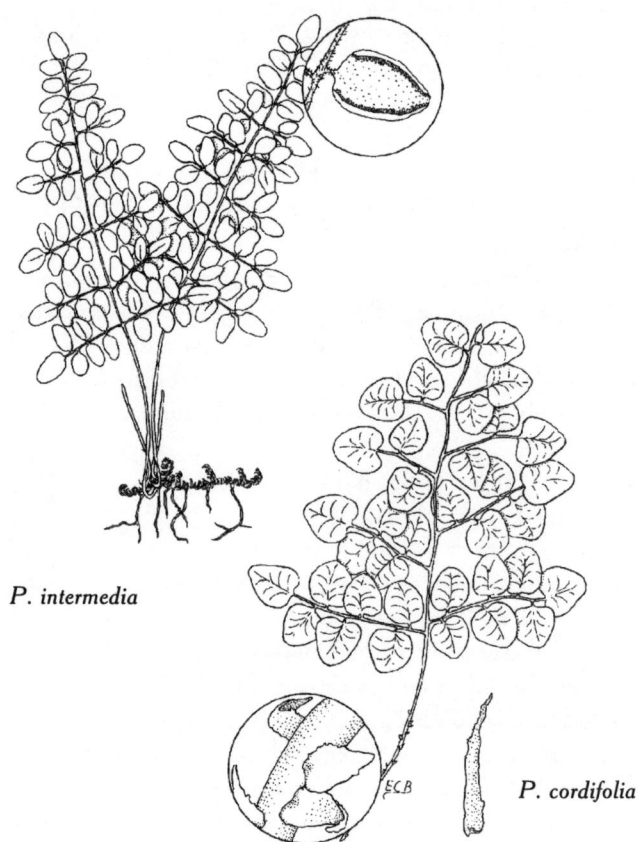

P. intermedia

P. cordifolia

Fig. 23. *Pellaea intermedia*, habit; pinnule. *P. cordifolia*, frond; stipe with scales; rhizome scale.

Pellaea species. In some localities, the leaves of zigzag cliff brake sprawl laxly and are supported by surrounding vegetation.

B   5. *Pellaea ternifolia* (Cav.) Link, TRANS-PECOS CLIFF BRAKE, TERNATE CLIFF BRAKE (*tern* = three + *folia* = leaf). Fig. 24. [*Pteris ternifolia* Cav.]. Rare on rocky cliffs, slopes, and ledges in igneous soil, in the Chisos Mts., Brewster 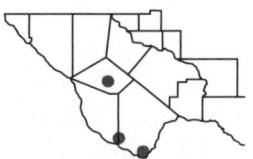 Co.; Big Bend Ranch State Park, Presidio and Brewster Counties; reported from the Davis Mts., Jeff Davis, Co. (Windham, 1993a). 3,500–7,000 ft. AZ, HI. Mexico. Central America. South America.

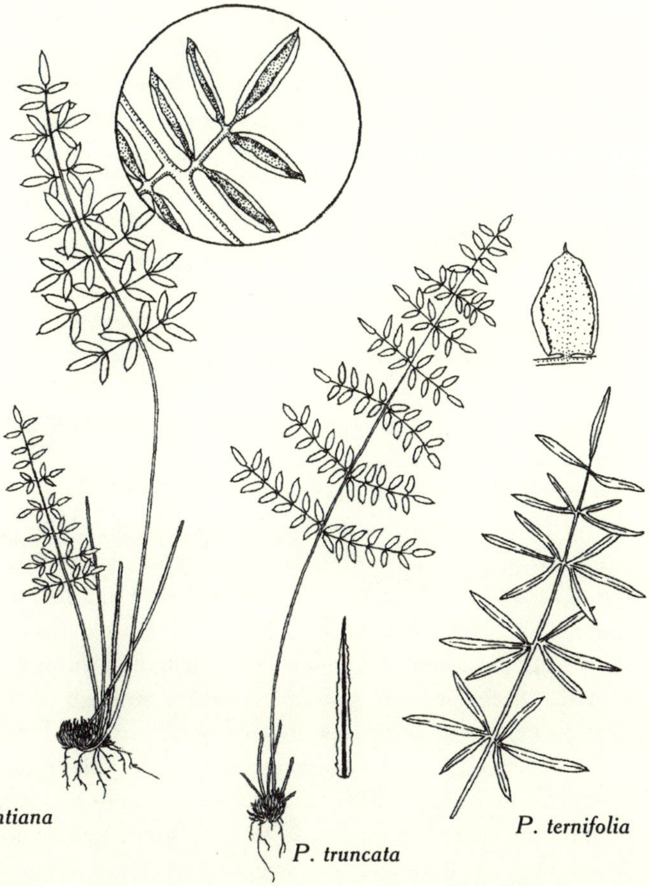

Fig. 24. *Pellaea wrightiana*, habit; pinnule. *P. truncata*, habit; pinnule; rhizome scale. *P. ternifolia*, blade.

According to Windham (1993b), three subspecies of *P. ternifolia* are separated by chromosome number and subtle morphological differences. The deeply pinnate-pinnatifid blades of *P. ternifolia* distinguish this species from the often similar *P. wrightiana*, which has truly pinnate blades.

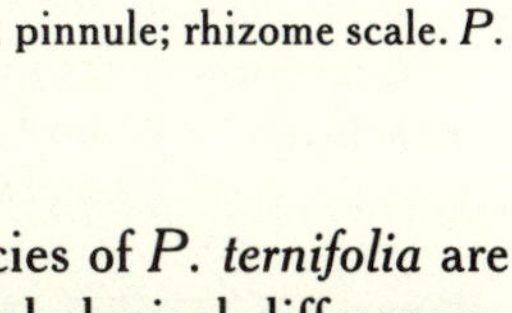

6. *Pellaea truncata* Goodd., SPINY CLIFF BRAKE (*trunc* = cut off, referring to the base of the pinnules). Fig. 24. [*P. longimucronata* Hook.]. Frequent to infrequent on rocky slopes and arroyos on limestone, igneous, or sandstone substrates in the Franklin and Hueco Mts., El Paso Co.; and Quitman

Mts., Hudspeth Co.; also reported (Correll, 1955) from Val Verde Co. 4,000–5,500 ft. NM, AZ, CO, UT, NV, CA. Mexico.

The fronds of *P. truncata* are almost entirely bipinnate, usually with 6–10 or more sets of pinnules. In contrast, the pinnae of *P. wrightiana* are generally pinnate-pinnatifid toward the top of the blade. The narrowly oblong pinnules of *P. truncata* have a truncate base, strongly inrolled edges, and a mucronate (short, sharp, abrupt) tip or spine at the apex. *Pellaea truncata* is more prevalent in the states to the west of the Trans-Pecos.

B,G 7. *Pellaea wrightiana* Hook., WRIGHT CLIFF BRAKE (for noted botanical collector, C. W. Wright). Fig. 24. [*P. ternifolia* var. *wrightiana* (Hook.) A. F. Tryon]. Infrequent in canyons and crevices, on rocky slopes, and on ledges, 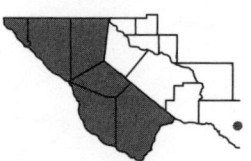 usually on igneous substrates but occasionally on limestone or Caballos Novaculite (S of Marathon, Brewster Co.), in El Paso, Culberson, Jeff Davis, Presidio, Brewster, and Val Verde Counties; also reported from Hudspeth Co. (Correll, 1955). (1,200) 3,800–7,000 ft. Most of Texas, except for Gulf Coast. NM, AZ, CO, UT, OK, NC. Mexico.

The linear-oblong blades of *P. wrightiana* are usually bipinnate toward the base with two, three, or more pairs of pinnules on each of the lower pinnae. Pinnation generally decreases upward, with pinnae often becoming trifoliolate or ternate toward the blade apex (fig. 24). The pinnules are elliptic to linear-oblong in shape and at first glance, the general aspect of *P. wrightiana* is similar to that of *P. ternifolia*. Upon closer examination, however, one can see that the pinnae of *P. ternifolia* are thrice divided to the base of the blade and not at all pinnate. Immature, sterile fronds of *P. wrightiana* can be deceptive and difficult to identify. Isozyme studies have demonstrated that *P. wrightiana* is a fertile tetraploid derived from *P. truncata* and *P. ternifolia*.

6. **Dennstaedtiaceae** Ching, Dennstaedtia Family

Plants of the Dennstaedtia family are usually terrestrial and often found in moist habitats. Their stems are rhizomatous, long- to short-creeping, and pubescent and occasionally bear scales. The leaf blades are simple to 4-pinnate, glabrous, or variously pubescent. Sori are marginal or submarginal and true indusia, which are cup- or pouch-shaped, are usually present. Gametophytes are green and cordate.

The family Dennstaedtiaceae is named for A. W. Dennstaedt, a noted German botanist. In the family there are about 20 genera and some 400 species that occur worldwide, usually in tropical regions. Four genera and six species occur in the United States, two genera and four species in Texas, and two species in separate genera are found in the Trans-Pecos.

Key to the Genera

1. Plants generally over 2 m tall; blades ovate in shape; pinnules with jointed reddish hairs on the abaxial veins; sori distinct, globose (spherical), not forming a continuous band along margins of pinnules . . . . 1. *Dennstaedtia*.
1. Plants generally under 2 m tall; blades broadly triangular in shape; pinnules densely pubescent abaxially with lax, spreading hairs; sori forming a continuous band along margins of pinnules . . . . 2. *Pteridium*.

1. *DENNSTAEDTIA* Bernh., CUPLET FERN

Plants of the genus *Dennstaedtia* are terrestrial and often form colonies. The stems are rhizomatous, long-creeping, and pubescent. The leaves of the Trans-Pecos species are generally 2–3 m long with blades that are ovate to deltate and 3–4-pinnate. The rachis, costae, and abaxial pinnule veins are pubescent with jointed, reddish hairs. The pinnules are lanceolate to ovate and glabrous above, with lobed margins. The sori are distinct, marginal, and globose.

There are about 70 species of *Dennstaedtia* distributed worldwide, with three species being found in the United States. Most members of the genus grow in wet tropical forests. However, one of the most common ferns of eastern moist temperate woods and meadows is *D. punctilobula*, the hay-scented fern, which is usually less than 1 m in height. A taller member of the genus, *D. bipinnata*, which can reach over 1.3 m high, is found only in a few locations in Florida. The tallest cuplet fern in the United States is found in the Trans-Pecos.

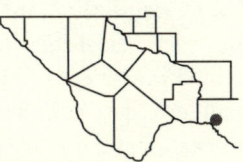

1. *Dennstaedtia globulifera* (Poir.) Hieron., BEADED CUPLET FERN (*glob* = a ball or globe, referring to the shape of the sori). Fig. 25. [*Polypodium globuliferum* Poir.]. Extremely rare in moist caves or limestone sinks in Val Verde Co. 1,800 ft. Mexico. West Indies. Central America. South America.

Fig. 25. *Dennstaedtia globulifera*, pinna; pinnule; sorus.

Dennstaedtia globulifera is one of the largest ferns in North America, re-portedly growing to 3 m and more in height. The only place where beaded cuplet fern occurs in the United States is in Val Verde Co., Texas.

2. *PTERIDIUM* Gled. ex Scop., BRACKEN FERN

Members of the genus *Pteridium* are terrestrial and colony forming and may grow to be 1 m or more in height. The Trans-Pecos *Pteridium* has hairy stems that are rhizomatous and long-creeping with leaves scattered along the rhizomes. Blades are coarse, broadly triangular, and 3-pinnate to 3-pinnate-pinnatifid. The undersides of the pinnae are densely covered with lax, contorted (twisted), spreading hairs. The sori generally form a continuous band at the pinnule margins.

According to Cranfill (1993), the genus *Pteridium* contains one spe-cies, *P. aquilinum*, and 12 varieties that are distributed nearly worldwide. Four varieties of this species are found in the United States, one occurring in the Trans-Pecos. The name is from the Greek *pteridion*, which means a small fern. The coiled young croziers or fiddleheads of *Pteridium* have long been eaten as a dietary staple in some cultures. They are especially popular in Japan, where fiddleheads are often dried for winter use or to make "warabi starch" for confections (Harrington, 1967). Recent research has revealed that *Pteridium* is mutagenic and carcinogenic. Although some of the toxicity is removed by cooking, *Pteridium* should not be eaten under any circumstances. Western bracken fern, with its long-creeping hairy rhizomes that grow deep in the soil, can be invasive in some areas, entering disturbed sites as an aggressive weed.

B   1. *Pteridium aquilinum* (L.) Kuhn var. *pubescens* Underw., DOWNY BRACKEN FERN, WESTERN BRACKEN FERN (*aquil* = eagle, referring to the resemblance of the fiddleheads to an eagle's talons; var. *pubesc* = downy). Fig. 26.
Found in the Trans-Pecos and in Texas on Mt. Livermore, Davis Mts., Jeff Davis Co.; also listed from the Rosillos Mts., Brewster Co. 6,900–7,500 ft. W to CA, WA, OR. N to Alberta, British Columbia, Canada.

   Pteridium aquilinum var. *pubescens* is the common bracken fern of the western states. It grows on open wooded slopes, in disturbed areas, and

Fig. 26. *Pteridium aquilinum* var. *pubescens*, habit; pinna; pinnules.

near streams. Western bracken fern is reported to be poisonous to livestock when eaten in large quantities. The best character for distinguishing the western variety of *Pteridium* from the three varieties occurring in more easterly locales is the abundant, spreading, lax pubescence on the underside of the pinnae.

7. **Thelypteridaceae** Ching ex Pic. Serm., Marsh Fern Family

Members of the family Thelypteridaceae may grow either on rock or in soil. The scaly stems are rhizomatous and creeping to ascending. The leaves are usually monomorphic, with blades that are pinnate-pinnatifid or 1–2-pinnate. The rachis of the blades may be grooved adaxially or not. Blades are occasionally glabrous but usually they are pubescent with needlelike, stellate, hooked, or septate (partioned, multicellular) hairs. Scales are generally absent. The sori are round, oblong, or elongate and inframedial to submarginal. Kidney-shaped (reniform) indusia are usually present. Gametophytes are cordate, green, and most often pubescent.

This large and complex fern family is treated by various authors as consisting of from one to 30 genera. Although members of the Thelypteridaceae have often been included in the Dryopteridaceae, there is no close relationship between the two families. The families differ in pubescence, presence or absence of blade scales, degree of pinnation, stipe characters, and vascular bundles as well as in chromosome number. The name is from the Greek *thelys*, meaning female, and *pteris*, for fern. As many as 30 genera and approximately 900 species occur worldwide, mostly in tropical areas. Of these, 1–3 genera and 25 species are found in the United States, two genera and six species occurring in Texas.

1. *THELYPTERIS* Schmidel, SHIELD FERN, FEMALE FERN, MAIDEN FERN

The Trans-Pecos ferns of the genus *Thelypteris* have stems that generally are long-creeping and leaves that are from 30 to 150 cm long. The leaves are pinnate-pinnatifid with abaxial pubescence of needlelike, transparent hairs. Adaxially the surfaces may be glabrous or pubescent with sparse, minute hairs. Scales are few, narrow, and abaxial on the costae. The round sori occur along the veins supramedially to submarginally. The indusia are usually kidney-shaped.

About 875 species are reported nearly worldwide, with 21 of these occurring in the United States, five in Texas, and one in the Trans-Pecos.

Thelypteris and *Adiantum capillus-veneris* are found in similar moist habitats and the two ferns often grow together.

B   1. *Thelypteris ovata* R. F. St. John var. *lindheimeri* (C. Chr.) A. R. Sm., LINDHEIMER MAIDEN FERN (*ovat* = eggshaped, for the shape of the blade; var. for F. J. Lindheimer, noted Texas plant collector). Fig. 27. [*Dryopteris*  *normalis* C. Chr. var. *lindheimeri* C. Chr.]. Infrequent in moist habitats in canyons, near springs, streams, and rivers, Terrell and Val Verde Counties; the lower canyons of the Rio Grande and one isolated location at a spring in Big Bend National Park, Brewster Co. 990–4,000 ft. Edwards Plateau, TX. Mexico. West Indies. Central America.

The pinnate-pinnatifid, ovate blades, submarginal sori, and needlelike hairs beneath the pinnae are the best characters for distinguishing *T. ovata*. The variety *ovata*, which has glabrous upper pinnae surfaces and lacks costal scales, occurs in the southeastern United States. *Thelypteris ovata* var. *lindheimeri*, the Trans-Pecos entity, was previously recognized as *Dryopteris normalis* (*Thelypteris normalis*), which is presently regarded as a synonym of *T. kunthii*.

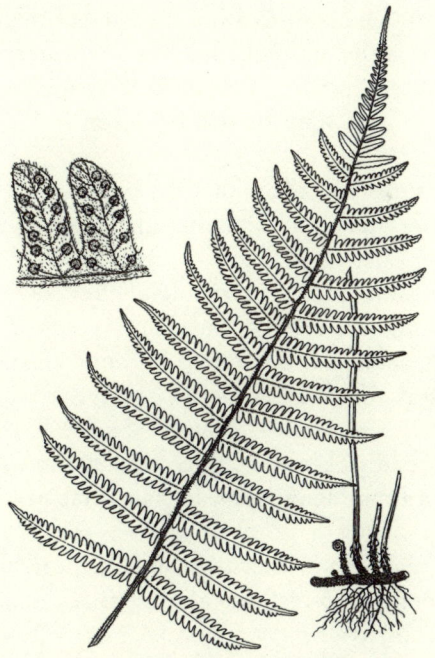

Fig. 27. *Thelypteris ovata* var. *lindheimeri*, habit; pinnules.

8. **Aspleniaceae** Newman, Spleenwort Family

Plants of the Aspleniaceae usually are found growing on rock or in soil. Occasionally tropical species are epiphytic. The stems are rhizomatous, generally erect, and occasionally long- or short-creeping, with latticelike clathrate scales. The leaf blades are simple to 4-pinnate and often pubescent with small glandular hairs and sparse, linear, clathrate scales. The sori, which occur on the veins, are linear or lunate (crescent-shaped). The indusia originate along one side of the sori and conform to their shape. The gametophytes are green and cordate.

The family Aspleniaceae consists of one large, diverse, worldwide genus of about 700 species. Approximately 30 species of *Asplenium* occur in the United States, five of these being found in Texas and four in the Trans-Pecos.

1. *ASPLENIUM* L., SPLEENWORT

Members of the genus *Asplenium* are most easily recognized by the clathrate stem scales and linear sori, which are arranged oblique to the midrib. Aspleniums usually grow in moist, shaded areas. The name *Asplenium* comes from the Greek *splen*, for spleen, denoting the ancient medicinal use of the genus in treating diseases of the spleen. *Asplenium* species hybridize readily, and polyploidy is widespread in the genus.

Key to the Species

1. Blades appearing like a tuft of grass . . . 3. *A. septentrionale.*
1. Blades fernlike, not appearing like a tuft of grass. . . . . (2).

2 (1). Apex of some blades becoming a whiplike extension
     terminating in a rooting bud . . . . 1. *A. palmeri.*
  2. Apex of blades not becoming a whiplike extension with a
     terminal rooting bud . . . . (3).

3 (2). Rachis reddish brown, pinnae usually oval, generally
     less than 6 mm long and less than twice as long as
     wide . . . . 4. *A. trichomanes.*
  3. Rachis black, pinnae usually oblong, generally more than
     6 mm long and 3–4 times as long as wide . . . 2. *A. resiliens.*

1. *Asplenium palmeri* Maxon, PALMER
SPLEENWORT (for E. J. Palmer, noted plant col-
lector). Fig. 28. Rare in crevices and on moist
shaded ledges of granite cliffs in the Davis Mts.,
Jeff Davis Co. 4,700–7,800 ft. Edwards Plateau,
TX, NM, AZ. Mexico. Central America.

Asplenium palmeri is unusual in having a long, naked, and whiplike (flagelliform) apex on some leaf blades. The flagelliform apex terminates in a
bud that produces a new plant. The margins of the pinnae of *A. palmeri* are
usually serrate (toothed) with sharply acute, distinct serrations.

B,G 2. *Asplenium resiliens* Kunze, BLACKSTEM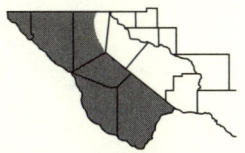
SPLEENWORT, LITTLE EBONY SPLEENWORT
(*resiliens* = recoiling). Fig. 28. Infrequent to frequent on woodland slopes, in shaded crevices, on
cliffs, and near boulders and springs in the moun-
tains of El Paso, Hudspeth, Culberson, Jeff Davis, Presidio, and Brewster
Counties, usually on igneous but occasionally on limestone substrates.
(1,700) 4,500–8,000 ft. C TX. Throughout much of the S U.S. Mexico.
West Indies. Central America. South America.

Asplenium resiliens is the most common of the Trans-Pecos spleenworts
and is usually found growing on or near rocks in moist, shady spots. The
name *resiliens* and the common names of blackstem and little ebony
spleenwort refer to the springy, black, lustrous stipe and rachis of this attractive fern.

B 3. *Asplenium septentrionale* (L.) Hoffm.,
FORKED SPLEENWORT (*septentrion* = northern).
Fig. 29. In crevices of noncalcareous ledges and
rocks. One collection from Boot Spring, Chisos
Mts., Brewster Co. ca. 7,000 ft. W U.S. Eurasia.
Asia.

Asplenium septentrionale bears a strong resemblence to a tuft of grass
and is easily overlooked by all but the most observant eyes. The narrow
leaves are less than 15 cm long, and the stipe, which is longer than the
blade, is often forked.

4. *Asplenium trichomanes* L. ssp. *tricho-*
manes, MAIDENHAIR SPLEENWORT (*tricho* =
hairy + *manes* = a cup). Fig. 28. Reported in
Texas (Correll, 1955) only from sheltered

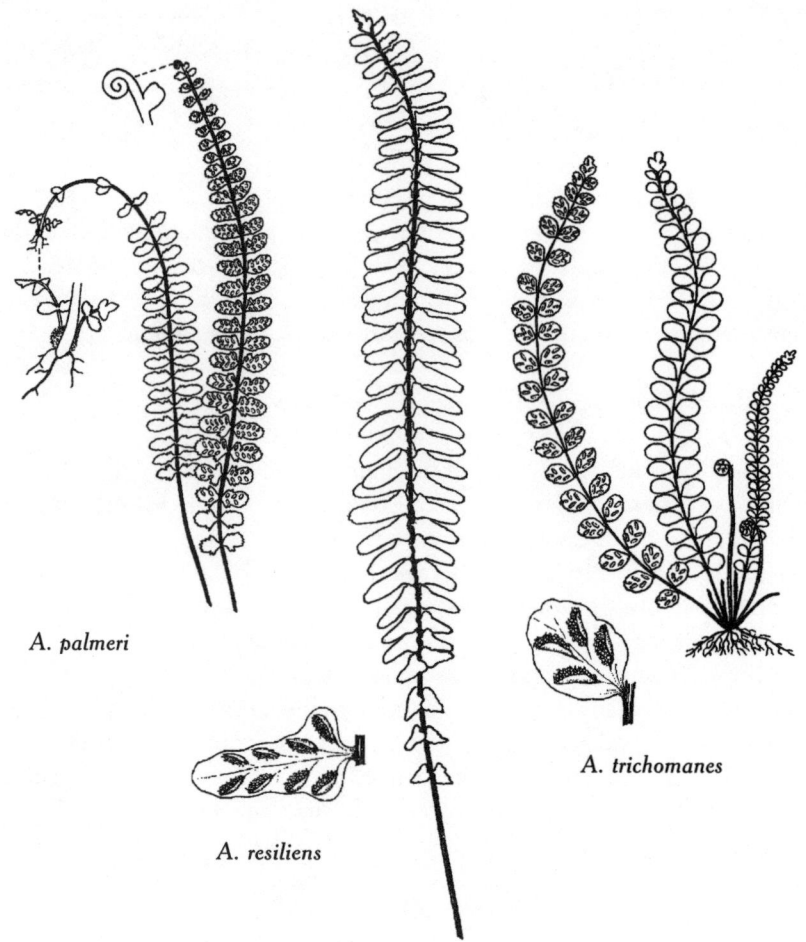

A. palmeri

A. trichomanes

A. resiliens

Fig. 28. *Asplenium palmeri*, fronds; rooting frond tips. *A. resiliens*, frond; fertile pinna. *A. trichomanes*, habit; fertile pinna.

crevices of ledges and cliffs, Mt. Livermore and Sawtooth Mt., Davis Mts., Jeff Davis Co. 6,500–7,500 ft. NM, AZ, CO, WY, OK, SD, far W states and Canada, E U.S. except FL. Eurasia. Asia. Africa. Australia.

The common name of *A. trichomanes* is maidenhair spleenwort because the round pinnae of the species are reminiscent of those of maidenhair fern, *Adiantum capillus-veneris*. The pinnae of two other Trans-Pecos spleenworts, *A. palmeri* and *A. resiliens*, tend to be more oblong in shape. *Asplenium trichomanes* ssp. *trichomanes*, which occurs mostly on acidic rocks such as basalt and granite, is known to occur at only a few sites in the

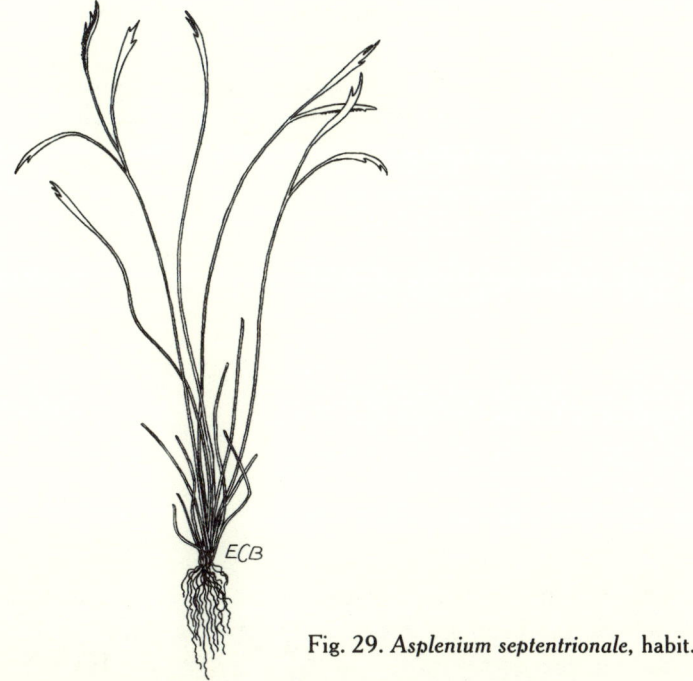

Fig. 29. *Asplenium septentrionale*, habit.

Trans-Pecos. *Asplenium trichomanes* ssp. *quadrivalens*—which differs from ssp. *trichomanes* mainly in chromosome number and occurs commonly on calcareous substrates in the northeastern and northwestern United States—is expected but has not been documented to occur in the Trans-Pecos. Maidenhair spleenwort is an excellent small fern for rock gardens.

9. **Dryopteridaceae** Herter, Wood Fern Family

Members of the Dryopteridaceae can usually be found growing on rock or in soil. Occasionally some species are epiphytic. The stems of the wood fern family are rhizomatous, creeping to erect, and they have scales, as do the petioles. The leaf blades are simple to 5-pinnate and glandular; hairs and scales are often present. The sori may either be discrete (separate) or may cover the abaxial surfaces. They are variously shaped and occur on the veins or at the vein tips. Generally they are not marginal. Indusia are usually present and range from round to oblong or elongate in shape. The gametophytes are green, cordate, and either pubescent or glabrous.

The name of the family comes from the Greek *drys* for tree and *pteris* for fern. About 60 genera and more than 3,000 species occur worldwide, with 18 genera and 79 species in the United States. Eight genera and 17 species occur in Texas, and four genera and 10 species are found in the Trans-Pecos.

Key to the Genera

1. Indusia cuplike, filaments or scalelike segments completely
   surrounding the sorus . . . . 4. *Woodsia.*
1. Indusia centrally or laterally attached, not cuplike . . . . (2).

2(1). Indusia round, attached at the center (peltate)
   . . . . 3. *Phanerophlebia.*
2. Indusia round-reniform (kidney-shaped) or hoodlike,
   attached laterally . . . . (3).

3(2). Indusia round-reniform, conspicuous . . . . . . 2. *Dryopteris.*
3. Indusia hoodlike, arching over the sori (may be obscure
   on mature leaves) . . . . 1. *Cystopteris.*

1. *CYSTOPTERIS* Bernh., BLADDER FERN, BRITTLE FERN, FRAGILE FERN

Plants of the genus *Cystopteris* are found growing on rock or in soil. The stems are rhizomatous, short- to long-creeping, having scales that are often clathrate. Leaf blades are ovate-lanceolate to deltate and 1–3-pinnate-pinnatifid, with the apex being pinnatifid. The pinnae are glabrous adaxially and pubescent or glabrous abaxially. Round sori occur in one row between the pinnule midrib and margin. The hoodlike indusia, which arch over the sori toward the margin of the segment, are often absent at maturity.

Cystopteris species hybridize readily where their ranges overlap. The hybrids usually have malformed or shriveled spores. The name *Cystopteris* comes from the Greek *kystos* for bladder and *pteris* for fern, referring to the indusium, which is inflated when young. Recent studies have shown that *C. fragilis*, which is widespread in North America, is a complex of several species and their fertile and sterile hybrids. About 20 species of *Cystopteris* are found worldwide, with nine occurring in the United States, three in Texas, and three in the Trans-Pecos.

Key to the Species

1. Rachis, costae, midribs, and indusia mostly without glandular hairs (some may be present in the axils on the pinnae); blade shape lanceolate to elliptic, usually widest at the middle or just below; bulblets never present on the rachis or costae . . . . 2. *C. reevesiana*.
1. Rachis, costae, midribs, and indusia with sparse to dense glandular hairs; blade shape ovate to deltate, usually widest at the base; bulblets often present on the rachis or costae. (2).

2 (1). Rachis with dense glandular hairs, bulblets often present; blades broad- to narrow-deltate; apex long-attenuate . . . . 1. *C. bulbifera*.
    2. Rachis with sparse glandular hairs, occasionally with misshapen bulblets; blades deltate to narrow-deltate; apex short-attenuate . . . . 3. *C. utahensis*.

G   1. *Cystopteris bulbifera* (L.) Bernh., BULBLET BLADDER FERN (*bulb* + *fer* = carry). Fig. 30. [*Polypodium bulbiferum* L.]. Ledges and crevices on moist limestone cliffs, often near waterfalls and streams in the Guadalupe Mts.,  Culberson Co. 5,500–6,500 ft. NM, AZ, UT. E and C U.S. SE Canada.

Cystopteris bulbifera has dense gland-tipped hairs on the rachis, costae, and indusia. Mature specimens often have deciduous round, pealike bulblets, which fall off very easily, on the rachis and costae. The blades are usually narrow-deltate with long attenuate apexes.

G   2. *Cystopteris reevesiana* Lellinger, SOUTH-WESTERN BRITTLE FERN (for T. Reeves, noted fern botanist). [*C. fragilis* var. *tenuifolia* (Clute) Broun]. On rock or in soil in McKittrick Canyon, Guadalupe Mts., Culberson Co., and Mt. Liver- 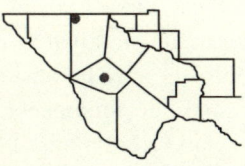 more, Davis Mts., Jeff Davis Co., on limestone or igneous substrates. 5,500–8,000 ft. NM, AZ, CO, UT. Mexico.

Cystopteris reevesiana does not have bulblets or gland-tipped hairs on the rachis, costae, or indusia. However, occcasionally a few gland-tipped hairs may be present in the axils of the pinnae.

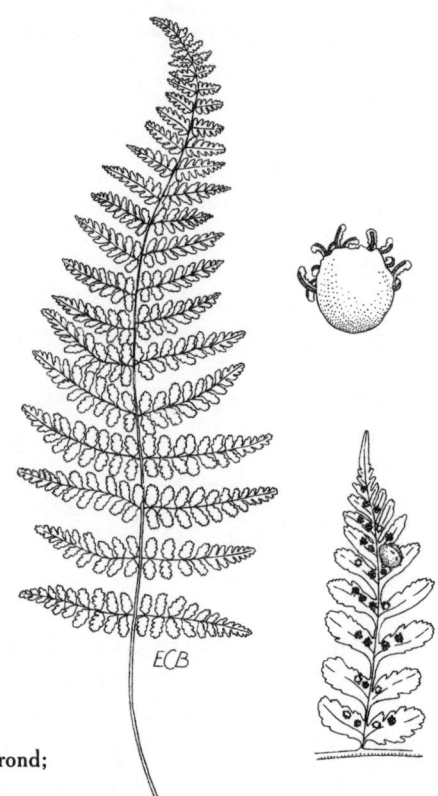

Fig. 30. *Cystopteris bulbifera,* frond;
sorus; fertile pinna with bulblet.

G   3. *Cystopteris utahensis* Windham &
Haufler, UTAH BLADDER FERN (for the type
locality in the state of UT). Cracks and ledges on
limestone cliffs in Culberson Co. Ca. 6,800 ft.
AZ, CO, UT.

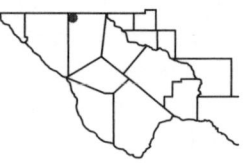

    Cystopteris utahensis displays some characters that are intermediate be-
tween its putative (probable) parents, C. *bulbifera* and C. *reevesiana.*
Bulblets may be absent or present and, if present, are often misshapen.
Glandular pubescence on the rachis, costae, and indusia is usually sparser
than in C. *bulbifera.* The three Trans-Pecos *Cystopteris* species are similar
in general appearance and best differentiated by their key characters and
location.

2. *DRYOPTERIS* Adans., WOOD FERN

Plants of the genus *Dryopteris* usually grow in soil, but occasionally they may be found growing on rock. The stems are rhizomatous and short-creeping to erect. The leaf blades are 1–3-pinnate-pinnatifid and deltate-ovate to lanceolate in shape. Usually the blades are glabrous adaxially. Sometimes scales or glands are found on the abaxial surface. The round sori are in one row between the midribs and margins. The round-reniform indusia are attached on one side at a sinus.

The name *Dryopteris* comes from the Greek *drys*, for tree, and *pteris*, for fern. Many species of *Dryopteris* are desirable garden plants. About 250 species are distributed mostly in the temperate areas of Asia, with four species being found in the United States. Two species occur in Texas; both are found in the Trans-Pecos.

Key to the Species

1. Blade 3-pinnate-pinnatifid at the base; petiole scales
   cinnamon-colored, of only one type . . . . 1. *D. cinnamomea.*
1. Blade pinnate-pinnatifid to 2-pinnate at the base;
   petiole scales light brown, of 2 kinds, one broad and
   one hairlike . . . . 2. *D. filix-mas.*

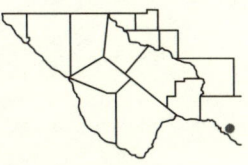

1. *Dryopteris cinnamomea* (Cav.) C. Chr., CINNAMON WOOD FERN (for the cinnamon-colored scales). Fig. 31. [*Tectaria cinnamomea* Cav.]. Rare in caves, Val Verde Co. 980 ft. AZ. Mexico.

The only documented occurrence of *D. cinnamomea* in Texas is from a cave near Comstock, Val Verde Co., in 1965. This rare fern also occurs on cliffs in S AZ and throughout central Mexico (Mickel, 1992).

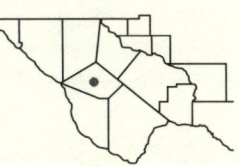

2. *Dryopteris filix-mas* (L.) Schott, MALE FERN [*filix* = fern + *mas* = male, referring to the species' vigorous nature). Fig. 32. [*Polypodium filix-mas* L.]. On ledges, bluffs, cliffs, and in moist woods in the Davis Mts., Jeff Davis Co. 6,000–8,000 ft. W and NE U.S. into Canada. Greenland. Europe. Asia.

Male fern is the only species of *Dryopteris* with two distinct kinds of petiole scales, one broad and scalelike and the other hairlike. *Dryopteris*

Fig. 31. *Dryopteris cinnamomea*, habit; pinnules.

filix-mas has been used medicinally, especially as a powerful vermifuge (to expel intestinal worms), from Greek antiquity to the present day. The spores of this species were thought to make the user invisible; as Shakespeare says in *Henry IV*, "We have the receipt of fern-seed, we walk invisible." Cool, moist, rocky woods in western and northeastern North America are the preferred habitat for *D. filix-mas*.

3. *PHANEROPHLEBIA* C. Presl, HOLLY FERN

Plants of the genus *Phanerophlebia* mainly grow in soil and are rarely found growing on rock. The stems are rhizomatous and short-creeping to erect. Leaf blades are ovate-lanceolate and once-pinnate. The pinnae grow

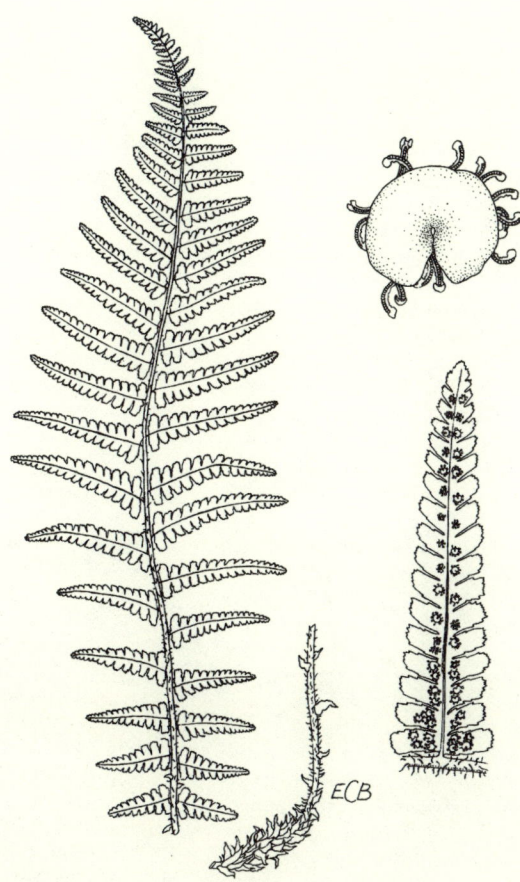

Fig. 32. *Dryopteris filix-mas*, habit; sorus; pinna.

with or without an eared lobe. The costae and veins have filiform (thread-like) scales abaxially and usually are glabrous adaxially. Margins are serrulate (with fine, sharp teeth) to spinulose (with small spines). The round sori grow in two or more rows. Indusia are round and peltate (attached at the middle) and persistent or disappearing at maturity.

The name *Phanerophlebia* is from the Greek *phaneros*, for free, and *phlebium* for vein, indicating the type of venation occurring in some species of the genus. Eight species occur in Mexico, Central America, South America, and the West Indies. Two species occur in the United States, Texas, and the Trans-Pecos.

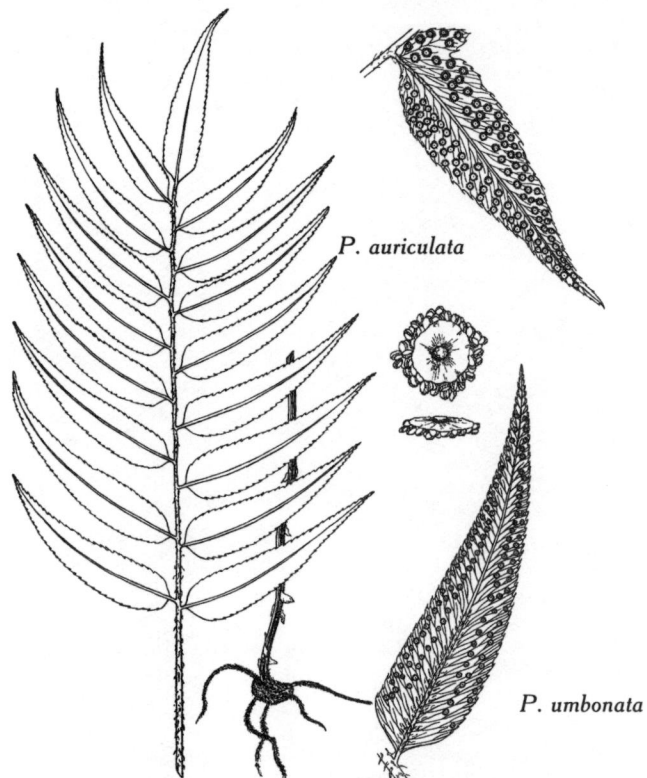

Fig. 33. *Phanerophlebia umbonata*, habit; pinna; sori. *P. auriculata*, pinna.

Key to the Species

1. Earlike lobes present, usually prominent on pinnae bases; indusia without a central, raised umbo (knob); indusia often disappearing at maturity. . . . . 1. *P. auriculata*.
1. Earlike lobes absent from the base of all pinnae; indusia with a central, raised umbo; indusia always present, not disappearing at maturity. . . . . 2. *P. umbonata*.

G   1. *Phanerophlebia auriculata* Underw., EARED HOLLY FERN (*auri* = ear + *cula* = little). Fig. 33. [*Cyrtomium auriculatum* (Underw.) C. V. Morton]. Rare in moist, sheltered areas in canyons in El Paso, Culberson, Jeff Davis, and Brewster Counties. 4,000–5,500 ft. NM, AZ. Mexico.

The most distinctive character of *P. auriculata* is the ear at the base of the pinnae, which occurs to a greater or lesser degree on at least some of the pinnae. The round, centrally connected indusia are obvious in younger specimens but disappear as the plant matures.

B   2. *Phanerophlebia umbonata* Underw., CHISOS HOLLY FERN (*umbo* = a projecting knob). Fig. 33. [*Cyrtomium umbonata* (Underw.) C. V. Morton]. Rare in moist spots in sheltered canyons in the Chisos Mts., Brewster Co. 5,000– 7,000 ft. Mexico.

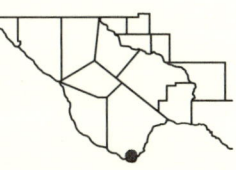

The only place in the United States where *P. umbonata* occurs is in the Chisos Mts. in Brewster County. The lack of ears on the pinnae of *P. umbonata* as well as the umbonate and persistant indusia distinguish this species from *P. auriculata*.

4. *WOODSIA* R. Brown, CLIFF FERN, WOOD FERN

Plants of the genus *Woodsia* are generally found growing on or among rocks. The stems are rhizomatous and creeping to compact, and they may be ascending to erect. The leaf blades are ovate, lanceolate or linear, and 1–2-pinnate-pinnatifid, usually with glandular hairs both abaxially and adaxially. The round sori are in one row between the margins and midribs. Indusia are persistent, with variously lobed, fringed, or scaled cups, which are basally attached around the sorus.

Woodsia is named for the English botanist, Joseph Woods. About 30 species occur in temperate and tropical regions, with 10 species found in the United States, five species in Texas, and three in the Trans-Pecos.

Key to the Species

1. Blades densely glandular, often viscid (sticky), lanceolate to ovate in shape; indusia segments usually broad at base, lower part of the segments consisting of several cells for most of their length. . . . . 3. *W. plummerae.*
1. Blades glabrous to sparsely glandular, never viscid, lanceolate to linear in shape; indusia segments narrow, filamentous (formed of threadlike filaments), consisting of only one cell for most of their length . . . . (2).

2 (1). Largest pinnae divided into 3–7 pairs of pinnules that are
closely spaced; pinnule margins with 1–2-celled short,
translucent projections; pinnule tips rounded to abruptly
tapered . . . . 1. *W. neomexicana.*
   2. Largest pinnae divided into 7–18 pairs of widely spaced
pinnules; pinnule margins with multicellular translucent
projections that often form twisted filaments; pinnule tips
narrowly acute to attenuate . . . . 2. *W. phillipsii.*

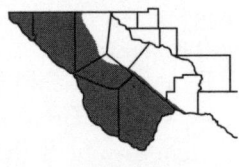

B,G 1. *Woodsia neomexicana* Windham, NEW
MEXICO CLIFF FERN (for the NM locality of the
type specimen). Fig. 34. Infrequent on igneous
cliffs and rocky slopes in mountainous areas of El
Paso, Hudspeth, Culberson, Jeff Davis, Presidio,
and Brewster Counties. 3,500–7,500 ft. NM, AZ, CO, OK, SD.

The taxa *W. neomexicana* and *W. phillipsii* have recently been separated (Windham, 1993b) from *W. mexicana,* which may also occur in the
Trans-Pecos. These three ferns, which are similar in appearance, may
have one parent in common.

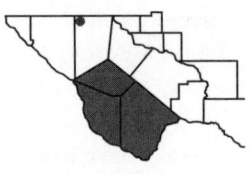

B,G 2. *Woodsia phillipsii* Windham, PHILLIPS
CLIFF FERN (for W. S. Phillips, noted fern botanist). Fig. 34. Infrequent, usually on igneous rocky
slopes and cliffs in mountainous areas of Culberson, Jeff Davis, Presidio, and Brewster Counties. 4,000–8,250 ft. NM, AZ. Mexico.

Woodsia phillipsii is difficult to distinguish with certainty from *W.
neomexicana* without magnification. Generally the larger number of widely
spaced pinnules of *W. phillipsii,* as opposed to the fewer, more closely
spaced pinnules of *W. neomexicana,* help to differentiate the two taxa.

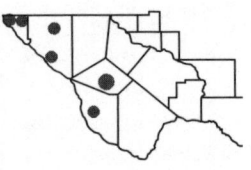

3. *Woodsia plummerae* Lemmon, PLUM-
MER WOODSIA, FLOWERCUP FERN (for Plum-
mer). Fig. 34. [*W. obtusa* (Spreng.) Torr. var.
glandulosa D. C. Eaton & Faxon; *W. obtusa* var.
plummerae (Lemmon) Maxon; *W. pusilla* E.
Fourn. var. *glandulosa* (D. C. Eaton & Faxon) T.
M. C. Taylor]. Infrequent on cliffs and rocky slopes on igneous or limestone substrates in mountainous areas of El Paso, Hudspeth, Jeff Davis,
and Presidio Counties. 4,000–8,000 ft. NM, AZ, CO, CA, OK.
Mexico.

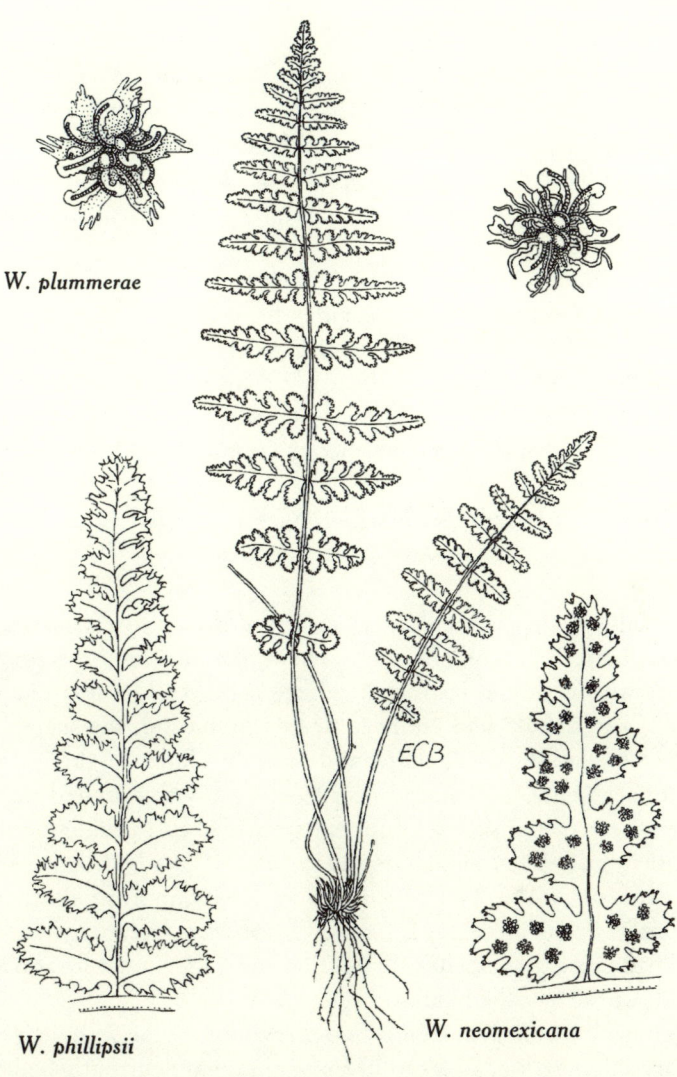

W. plummerae

W. phillipsii

ECB

W. neomexicana

Fig. 34. Woodsia phillipsii, pinna. W. plummerae, sorus. W. neomexicana, habit; sorus; pinna.

Woodsia plummerae is usually very glandular and viscid with indusial segments that are broad below the middle and filamentous only near the tips. Woodsia neomexicana and W. phillipsii are only sparsely glandular and have almost entirely filamentous indusia. Both W. plummerae and W. phillipsii may occasionally have bifurcate (forked) frond tips.

10. **Polypodiaceae** Bercht. & J. Presl, Polypody Family

Plants of the Polypodiaceae are found growing on rock or in soil. Occasionally some species may be epiphytic. The stems are short- to long-creeping and scales are present. The leaf blades are simple, pinnatifid to pinnate. Blade hairs and scales are present or absent. The sori, which are round to oblong, have no indusia. Gametophytes are green, cordate to elliptic, and glandular or glabrous.

The family name Polypodiaceae is from the Greek *poly,* for many, and *podion,* for little foot, alluding to the many-branched creeping rhizomes of some members of the family. About 40 genera and 500 species occur throughout the world, generally in the tropics and subtropics. Seven genera and 25 species are found in the United States. One genus with three species occurs in Texas and two of these species are found in the Trans-Pecos.

1. *PLEOPELTIS* Humb. & Bonpl. ex Willd., SHIELDED SORUS FERN

Plants of the genus *Pleopeltis,* which can grow up to 25 cm high, are usually epiphytic or found on rock. The stems are long-creeping with bicolored scales with clathrate (latticelike) centers. The leaf blades are linear-oblong to deltate and entire to deeply pinnatifid; the scales on the abaxial surfaces are stalked, peltate, and centrally clathrate; there are up to 25 pairs of segments, which are linear to oblong with rounded tips. The sori are discrete (separate) or confluent (joined), and usually occur in a row on either side of the midrib on the distal half of a leaf. Each sorus is covered, when immature, by a false indusium of overlapping peltate scales.

The name *Pleopeltis* comes from the Greek *pleos,* for many, and *pelte,* for shield, referring to the peltate scales that cover the immature sori as well as the undersurface of the blades.

The genus *Pleopeltis* is currently under revision. Some scaly-leaved taxa that had traditionally been placed in *Polypodium* are currently believed to be more closely related to *Pleopeltis* and have recently been transferred to that genus. About 50 species of the mainly epiphytic fern genus *Pleopeltis* occur in tropical regions of the world. Four taxa are found in the United States, three in Texas, and two in the Trans-Pecos.

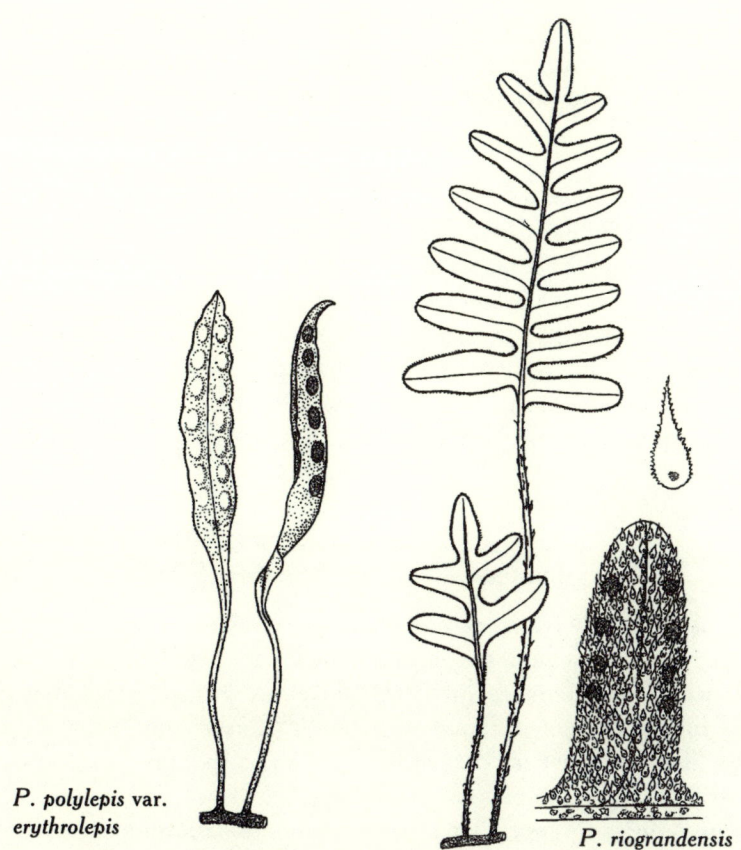

P. polylepis var.
erythrolepis

P. riograndensis

Fig. 35. *Pleopeltis polylepis* var. *erythrolepis*, habit. *P. riograndensis*, habit; pinnule; scale

Key to the Species

1. Blades entire-margined . . . 1. *P. polylepis* var. *erythrolepis*.
1. Blades deeply pinnatifid . . . . 2. *P. riograndensis*.

1. *Pleopeltis polylepis* (Roem. ex Kunze)
T. Moore var. *erythrolepis* (Weath.) T. Wendt,
RED SCALE POLYPODY (*poly* = many + *lepi* =
scale, referring to the dense, abaxial scales; var.
erythro = red). Fig. 35. [*Polypodium*
erythrolepis Weath.; *P. erythrolepis* (Weath.) Pic. Serm.]. Extremely
rare, upper Limpia Canyon, Davis Mts., Jeff Davis Co. 7,000 ft. Mexico.

The only place in the United States where *P. polylepis* var. *erythrolepis* is known to occur is at one site in upper Limpia Canyon in the Davis Mountains, where it has been reported to form extensive mats on porphyritic rocks.

B   2. *Pleopeltis riograndensis* (T. Wendt) E. G. Andrews & Windham, Rɪᴏ Gʀᴀɴᴅᴇ Sᴄᴀʟʏ Pᴏʟʏᴘᴏᴅʏ (for the Rio Grande). Fig. 35. [*Polypodium thyssanolepis* A. Braun ex Klotzsch var. *riograndensis* T. Wendt]. In canyons, on rocky 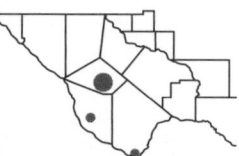 ledges, slopes, and in crevices in moist, shaded locales at Boot Spring, Chisos Mts., Brewster Co.; Davis Mts., Jeff Davis Co.; reported (Johnston, 1990) from the Chinati Mts., Presidio Co. 4,700–7,500 ft. AZ. Mexico.

11. **Marsileaceae** Mirb., Water Clover Family

Plants of the family Marsileaceae are found growing in or near water. They often form colonies and are usually rooted in mud, although they may also be floating. The stems are rhizomatous. Leaves are long-petioled, with expanded blades present in the genus *Marsilea* but absent in the genus *Pilularia*. The sori have bean- or peashaped sporocarps (fig. 36) on short stalks near the base of the petioles. Sporangia are of two types, which are housed within the same sporocarp. The megasporangia produce one spore and the microsporangia have up to 64 microspores.

There are three genera and about 50 species distributed nearly worldwide in temperate and tropical areas, with two genera and seven species found in the United States. Two genera occur in Texas and one of these is found in the Trans-Pecos.

1. *MARSILEA* L., WATER CLOVER, PEPPERWORT

Marsilea was named for Count Luigi Marsigli, a noted seventeenth-century Italian mycologist. One common name, water clover, comes from the resemblance of *Marsilea* to a four-leaf clover, because the blade is palmately divided into four pinnae. The other common name, pepperwort, is attributed to the resemblance of the sporocarps to peppercorns. The sporocarps, which are borne on stalks near the base of the petioles, are more or less pubescent, often have projecting teeth, and separate into two valves when they rupture just prior to fertilization. Presence of mature sporocarps is essential in identifying specimens with certainty.

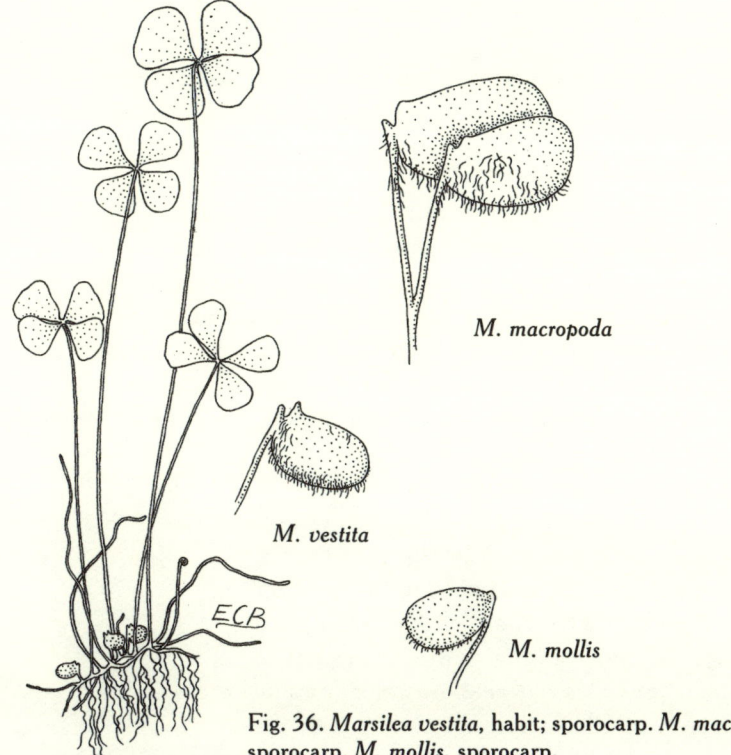

Fig. 36. *Marsilea vestita*, habit; sporocarp. *M. macropoda*, sporocarp. *M. mollis*, sporocarp.

About 45 species occur almost worldwide, with six being found in the United States, three in Texas, and three in the Trans-Pecos.

Key to the Species

1. Superior (topmost) tooth of sporocarp 0.4–1.2 mm in length; pinnae sparsely pubescent to glabrous abaxially   3. *M. vestita*.
1. Superior tooth of sporocarp absent or to 0.2 mm in length; pinnae decidedly pubescent abaxially. . . . . (2).

2 (1). Pinnae conspicuously hairy abaxially with long loose hairs; sporocarps strongly ascending, 6–9 mm in length; sporocarp peduncles usually branched with 2–6 sporocarps per stalk . . . . 1. *M. macropoda*.
2. Pinnae pilose (shaggy with soft, long hairs) abaxially; sporocarps nodding or perpendicular, 5 mm or less in length; sporocarp peduncles unbranched with 1 sporocarp per stalk. . . . . 2. *M. mollis*.

1. *Marsilea macropoda* Engelm. ex A.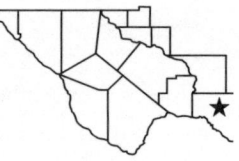
Braun, LARGE FOOT WATER CLOVER (*macro*
= large + *pod* = foot). Fig. 36. Reported
(Correll, 1955) from Val Verde Co. Ca. 1,200
ft. S-C TX, LA, AL. Mexico.

Marsilea macropoda is larger than the other Trans-Pecos water clovers
and has broad pinnae with spreading, white, abaxial hairs that show as an
edging when viewed from the adaxial side. The sporocarps, which virtually
lack a superior tooth, are borne on long, often branched peduncles. Large
foot water clover is used as a ground cover in the San Antonio Botanical
Garden in moist, shaded areas. A hybrid between *M. macropoda* and *M.
vestita*, having pinnules bordered with long hairs as in *M. macropoda* and
toothed sporocarps as in *M. vestita*, occasionally occurs in areas where the
two species overlap.

2. *Marsilea mollis* B. L. Rob. & Fernald,
MEXICAN WATER CLOVER (*molli* = soft). Fig.
36. One collection cited (Johnson, 1986) from
Limpia Canyon, Davis Mts., Jeff Davis Co.; may
be expected throughout the central mountain re-
gion of the Trans-Pecos in shallow water or muddy areas. 3,500–6,000 ft.
AZ. Mexico. South America.

The sporocarps of *M. mollis* are small, usually nodding, lack a superior
tooth, and are covered with spreading hairs. The sporocarp peduncles are
slender and wiry. The name *M. mexicana* A. Braun has been improperly
used for this taxon in the past. The character of reddish streaking on the
pinnae, formerly attributed to *M. mollis,* occurs on the floating leaves of
many other *Marsilea* as well. In Mexico *M. mollis* bears sporocarps from
March through December and perhaps throughout the year (Mickel,
1992).

3. *Marsilea vestita* Hook. & Grev.,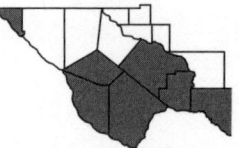
HOOKED WATER CLOVER (*vesti* = clothing, a
coat). Fig. 36. [*M. fournieri* C. Chr.; *M. mucronata* A. Braun; *M. tenuifolia* Engelm. ex A.
Braun; *M. uncinata* A. Braun; *M. vestita* ssp.
tenuifolia (Engelm. ex A. Braun) D. M. Johnson]. In ponds, wet depressions, stock tanks, and along streams and rivers in El Paso, Jeff Davis, Presidio, Brewster, Pecos, Terrell, and Val Verde Counties. 1,000–6,100 ft.
Throughout most of the W U.S. and S Canada. Mexico. South America.

Marsilea vestita exhibits a number of regional character variations across its large range. This has led to the use of many different names that are now considered synonymous. When *M. vestita* occurs in ephemeral water sources, the plants generally continue for 6–8 weeks after the water dries up and then disappear during the dry season (Felger, 2000). The most obvious identifying character of *M. vestita* is the conspicuous superior tooth of the sporocarp. *Marsilea mollis* and *M. macropoda* have sporocarps that are virtually without superior teeth. The hairs on the underside of the pinnae of *M. vestita* are appressed, while those of *M. mollis* are spreading and those of *M. macropoda* are long, white, and spreading.

12. **Azollaceae** Wettstein, Azolla Family

Plants of the Azollaceae are aquatic and generally found floating on still or slow-moving water or stranded on mud. The stems are branching and prostrate to nearly erect. The leaves are minute (0.6–2 mm wide), two-lobed, sessile, alternate, often imbricate, and two-ranked along the upper stem. The lower leaf lobe, which is submerged, is slightly larger than the upper lobe, and is usually translucent and colorless. The upper leaf lobe is emergent, green, and photosynthetic, or red when stressed, with a narrow and colorless margin. The two types of sporocarps are in pairs; megasporocarps contain one megasporangium and microsporocarps contain up to 130 microsporangia.

The single genus contains seven species that are distributed worldwide in tropical and temperate regions. Two species are found in Texas and one occurs in the Trans-Pecos.

1. *AZOLLA* Lam., MOSQUITO FERN

The species of *Azolla* are difficult to identify because the necessary distinguishing characters often are either absent or extremely difficult to recognize. Sori are lacking on most preserved specimens, and examination of megasporangial sculpturing and filaments requires the use of a scanning electron microscope. Much confusion about distribution of species has arisen because characters, especially those involving the glochidia (barbed hairs) that once were used for identification, have not proven to be consistent.

The genus name is from the Greek *azo*, to dry, and *ollyo*, to kill, alluding to the plants dying during dry spells. *Azolla* is noted for its symbiotic relationship with the nitrogen-fixing cyanobacterium *Anabaena azollae*. The

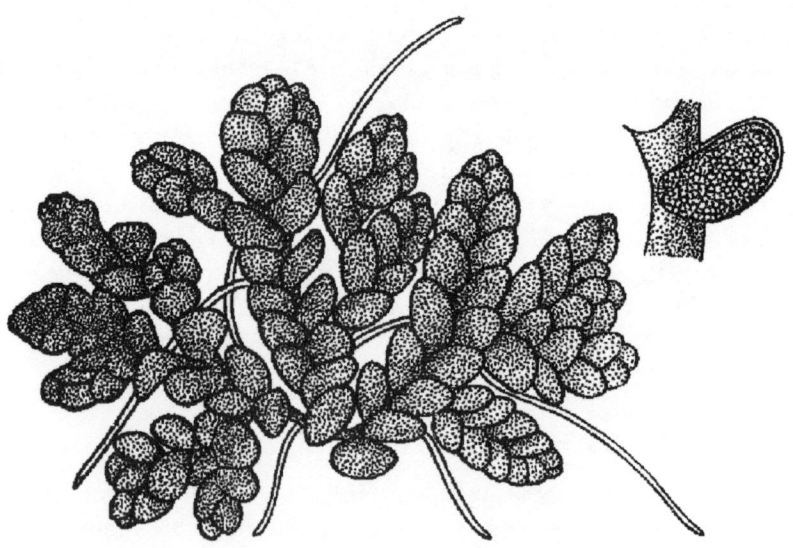

Fig. 37. *Azolla mexicana*, habit; leaf.

cyanobacterium occurs mostly in the cavities of the upper leaf lobes of *Azolla*, which allows the plant to be used as green fertilizer and as a nutritional animal food supplement. *Azolla pinnata* has long been grown as fertilizer in rice paddies in Asia, and several species of *Azolla* have been introduced into other rice-growing areas of the world as well. The economic importance of *Azolla* has made it the most frequently studied genus of ferns.

1. *Azolla mexicana* C. Presl, MOSQUITO FERN (for Mexico). Fig. 37. Rare to abundant in pools, springs, bogs, and streams in Brewster, Jeff Davis and Presidio Counties. 4,000–5,200 ft. Great plains, SW and W U.S. British Columbia, Canada. Mexico. Central America. South America.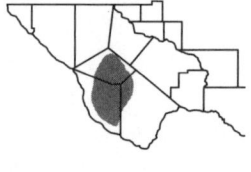

According to T. A. Lumpkin (1993) *A. mexicana*, a species generally found in western states, is the *Azolla* that occurs in the Trans-Pecos. This taxon is to be expected throughout the Trans-Pecos in still or slow-moving perennial waters. *Azolla caroliniana*, which is virtually indistinguishable vegetatively from *A. mexicana*, is typically a species of the eastern states, occurring in Texas to the east of the Pecos River.

Selected Glossary

Abaxial (compare adaxial). The side away from the axis; in ferns, the underside of a frond where sori may develop.

Abortive. Imperfectly developed; common in spores of hybrid ferns.

Abrupt. Changing suddenly rather than gradually.

Acuminate. Tapering to a long, drawn-out point.

Acute. Tapering to a short, sharp point.

Adaxial (compare abaxial). The side toward the axis; in ferns, the upper surface of a frond.

Adnate. Fused with a dissimilar structure.

Adventitious. Arising in an irregular or unusual place.

Aerial. Occurring above ground.

Agamospory. The overall asexual reproductive process involving both diplospory and apogamy.

Alternate (compare opposite). Borne singly at different levels on the axis or stem.

Amphibious. Growing equally well on land or in water.

Annual. Of one season's duration.

Annulus. A specialized ring of cells, partial or complete, along one side of the sporangium, involved in spore release.

Antheridium (pl. antheridia). Male sex organ on gametophytes of ferns and fern allies, producing male gametes (spermatozoids).

Apex. The tip or distal end of an organ.

Apical. At the apex.

Apiculate. Ending abruptly in a small, slender point.

Apogamy. Development of a sporophyte directly from the gametophyte (prothallus), without the occurrence of fertilization.

Apospory. The formation of a gametophyte (prothallus) directly on the sporophyte, without the production of spores.

Appressed. Closely pressed against something.

Aquatic. Growing wholly or partially submerged in water.

Archegonium (pl. archegonia). Female sex organ on gametophytes of ferns and fern allies, producing the female gamete (egg).

Areole. A surface discontinuity generally circular in outline.

Articulate. Jointed; with nodes or joints; with places where separation naturally takes place.

Ascending. Rising up; produced obliquely or indirectly upward.

Asexual reproduction. Reproduction without fusion of gametes; any vegetative reproduction, e.g., apogamy or by bulblets or bulbils.

Attenuate. Tapering long and gradually; drawn out.

Axil. The angle formed by a leaf or leaflet with the rhizome or rachis.

Axis. The main stem; main line of development.

Basifixed. Attached or fixed by the base.

Bicolored (compare concolored). Having two colors.

Bifurcate. Forked into two parts.

Bipinnate. Twice pinnately divided (fig. 3).

Bipinnate-pinnatifid. Twice pinnatifid (fig. 3).

Blade. The expanded part of a leaf or frond.

Bristle. A stiff hair.

Bulbil. A little bulb borne on a frond at the junction of main veins; developing into a plant.

Bulblet. A little bulb; same as a bulbil.

Capitate. Headlike.

Caudex. The persistent base of a perennial plant.

Cilia. Hairs forming a fringe, usually along the margin.

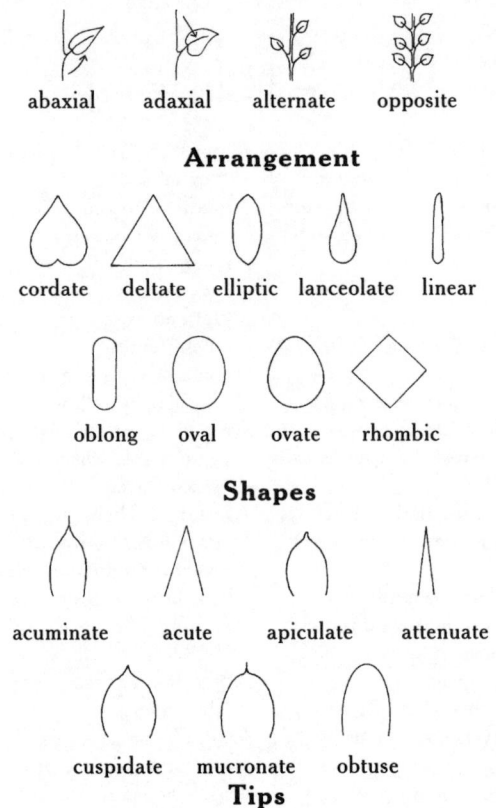

abaxial   adaxial   alternate   opposite

Arrangement

cordate   deltate   elliptic   lanceolate   linear

oblong   oval   ovate   rhombic

Shapes

acuminate   acute   apiculate   attenuate

cuspidate   mucronate   obtuse
Tips

Fig. 38. Glossary illustrations of arrangement, shapes, and tips.

Ciliate. With a marginal fringe of hairs.

Circinate. Coiled from the apex downward, as in young fern fronds.

Circinate vernation. Coiled or rolled lengthwise, from apex downwards, as in young fern fronds; coiled in bud.

Clathrate. With cell walls thickened in the form of a lattice.

Compound. Two or more similar parts of one organ, as in a frond with two or more leaflets.

Concolored (compare bicolored). Uniformly colored; a scale of only one color or both sides of a frond being the same color.

Confluent. Joined, running together.

Cone. Popular term for a strobilus.

Contorted. Twisted.

Cordate. Heart-shaped.

Costa (pl. costae). The midrib or rachis of a pinna.

Costal. Associated with the costa.

Creeping. Running along or under the ground, producing roots and shoots (or leaves) at intervals.

Crenate. Rounded, shallow-toothed; scalloped.

Crozier (alternate spelling, crosier). Coiled young frond; fiddlehead.

Cuneate. Wedge-shaped.

Cuspidate. Tapering gradually to a rigid tip.

Deciduous (compare persistent). Falling at maturity or at the end of the growing season.

Dehisce. Split open.

Dehiscence. The process of splitting open.

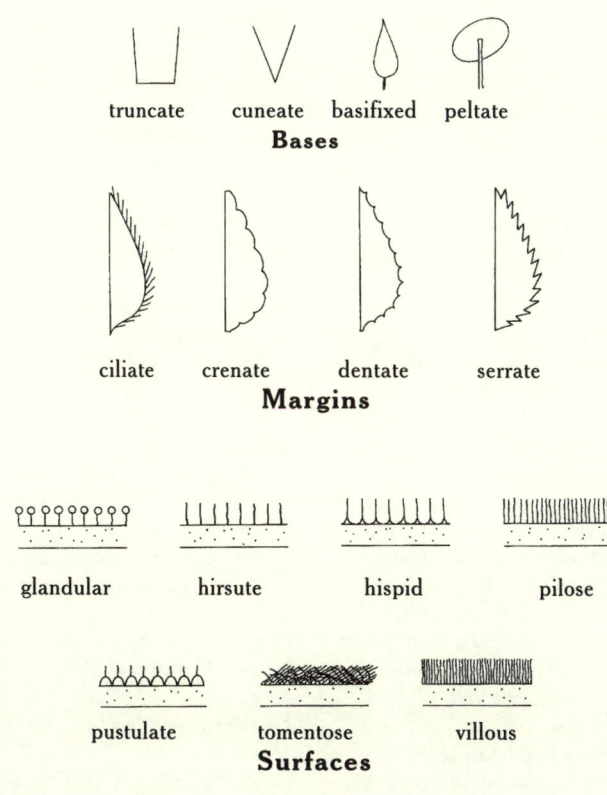

truncate   cuneate   basifixed   peltate

Bases

ciliate   crenate   dentate   serrate

Margins

glandular   hirsute   hispid   pilose

pustulate   tomentose   villous

Surfaces

Fig. 39. Glossary illustrations of bases, margins, and surfaces.

Deltate. Shaped like an equilateral triangle.

Deltoid. Same as deltate.

Dentate. With sharp teeth perpendicular to the margin.

Denticulate. Finely dentate.

Desiccate. Dry out.

Dichotomous. Forked in one or more pairs.

Dicot (see eudicot).

Dimorphic (compare monomorphic). Existing in two forms, having fronds of two types.

Diploid (compare haploid). Having two sets of chromosomes.

Diplospory. Production of unreduced spores that ultimately may give rise to unreduced gametophytes; a type of asexual (apomictic) reproduction.

Discrete. Separate, not fused or joined.

Dissected. Deeply divided into slender segments.

Distal (compare proximal). Toward the tip.

Divergent. Spreading broadly.

Divided. Separated to very near the base or midrib.

Dorsal (compare ventral). Pertaining to the back of an organ or the lower surface of a leaf.

Dorsiventral. With distinct upper and lower surfaces.

Elater. One of the four ribbonlike bands or filamentous appendages of the spores of *Equisetum* thought to be an aid to spore dispersal.

Elliptic. Narrowly oval in outline, widest at or near the middle, narrowing to rounded ends.

Embryo. A rudimentary plant, developing from a zygote.

Entire. Not lobed or divided; with a continuous margin (fig. 3).

Ephemeral. Lasting a short time.

Epiphyte. A plant that grows on another plant but does not draw food or water from it.

Erect. Upright.

Eudicot (compare monocot). Referring to a class of flowering plants, the eudicotyledons, that are distinguished by having two seed (embryo) leaves. Formerly known as dicot.

False indusium (pl. false indusia). A covering over the sorus formed by the recurved leaf margin, as in *Adiantum* and other members of the family Pteridaceae.

Farina. A mealy or flourlike waxy covering of a surface, as on the underside of fronds of *Notholaena* and *Argyrochosma*.

Farinose. Covered with farina.

Fern allies. Plants similar and probably related to ferns, reproducing by spores but having microphylls.

Fertile (compare sterile). Said of plants that produce spores or seeds.

Fiddlehead. A popular term for a fern crozier.

Filamentous. Formed of filaments.

Filiform. Threadlike.

Flagelliform. Long and whiplike.

Flora. The plant life of a given area; a book treating the plants of a region.

Free. Not joined to any other part.

Frond. The leaf of a fern or palm, including the blade and stipe or petiole.

Gamete. Sex cell, male or female; gametes unite at fertilization to form a zygote.

Gametophyte. The generation that bears the sex organs; in ferns the prothallus that bears antheridia and archegonia; contains half the number of sets of chromosomes that the sporophyte contains.

Glabrous (compare pubescent). Without hairs or scales.

Gland. A secreting part of a hair or appendage; a fluid-secreting organ.

Glandular. Bearing glands.

Globose. Spherical; globelike.

Glochidia. A barbed hair or bristle.

Grooved. Channeled or furrowed.

Habit. The general appearance of an organism.

Habitat. The environment in which a population lives.

Hair. A slender epidermal appendage, either unicellular or formed by a single column of cells.

Haploid (compare diploid). Having a single full set of chromosomes.

Herbaceous. Not woody, with soft tissue; dying down each year.

Heterosporous. Producing spores of two different kinds, microspores and megaspores.

Hirsute. With rather coarse, stiff hairs.

Hispid. With rough, firm, stiff, or bristly hairs.

Homosporous. Producing only one type of spore.

Hood. A hollow, arched covering over another structure or part.

Hoodlike. Conspicuously concave, like a hood.

Hybrid. The progeny resulting from a cross between two parents, most commonly used in reference to a cross between infraspecific, interspecific, or intergeneric taxa.

Imbricate. Overlapping, like shingles on a roof.

Indusium (pl. indusia). A thin epidermal outgrowth of tissue covering or partially covering the sorus on a fern leaf, at least when young.

Inframedial. Between the middle and the edge or margin.

Interspecific. Between species.

Intraspecific. Within a species.

Isozyme. A separable form of an enzyme.

Jointed. Bearing joints or nodes.

Juvenile. The young stage of growth.

Keel. A rather sharp, longitudinal ridge, like the bottom of a boat.

Lanceolate. Lance-shaped; much longer than it is wide, tapering to the apex, the widest part below the middle.

Lateral. On the side of an axis.

Lax. Loosely arranged.

Leaf-gap. A break in the vascular cylinder of a stem where the leaf-trace (or vascular bundle) arises.

Leaflet. A secondary segment of a compound leaf.

Linear. Long and narrow, the sides more or less parallel.

Lobe. A rounded division or segment of an organ; the lobe of a leaf.

Longitudinal (compare transverse). Parallel to the long or longitudinal axis.

Lunate. Crescent-shaped.

Lustrous. Glossy, shiny.

Margin. The edge.

Marginal. At the edge.

Mat. A tangled, low mass of stems.

Mealy. Covered with coarse granules or flourlike powder.

Medial. At or near the middle; between the midrib and margin.

Megaphyll. A leaf anatomically associated with a leaf-gap; a large complex leaf, such as a fern frond.

Megasporangium. A sporangium bearing megaspores, as in a heterosporous plant.

Megaspore. A large spore; in heterosporous plants, the larger spore that produces the female gametophyte and gametes.

Megasporocarp. A thick-walled organ containing at least one megasporangium.

Mesic. Characterized by a moderate amount of moisture.

Microphyll. A leaf anatomically without a leaf-gap; a sterile small leaf of a fern ally.

Microspore. A small spore; in heterosporous plants, the smaller spore that produces the male gametophyte and gametes.

Microsporocarp. A thick-walled organ containing microsporangia, as in *Marsilea*.

Midrib. The main vein of a leaf or leaflike part.

Monocot (compare eudicot). Referring to a class of flowering plants, the monocotyledons, that are distinguished by having one seed (embryo) leaf.

Monomorphic (compare dimorphic). Having fronds of one type.

Monostromatic. One cell layer thick.

Morphology. The form and structure of a plant.

Mosslike. Like a moss of the phylum Bryophyta.

Mucilaginous. With mucilage, a slimy material.

Mucronate. With a short, sharp, abrupt tip.

Nerve. A slender vein.

Net-veined. Reticulate, or netlike.

Nodding. Hanging down.

Non-circinate vernation. Hooked, rather than coiled, young fronds.

Obcordate. The reverse of cordate, with the broadest part at the apex.

Oblique. Slanting at an angle; with sides unequal.

Oblong. Longer than it is broad, with parallel sides.

Obtuse. Blunt or rounded at the apex.

Opposite (compare alternate). Arising on opposing sides at the same level.

Oval. Rounded, but longer than it is wide, broadly elliptic.

Ovate. Egg-shaped in outline.

Pectinate. Comblike or pinnatifid where deeply divided narrow parts are close together.

Pedate. Palmately divided with basal lobes again divided.

Peltate. Shieldlike; a stalked organ where attachment of the stalk is near the middle, or at least away from the margin.

Pendant. Hanging downward, or drooping.

Pentagonal. Five-sided.

Perennial. Living more than two years, often regenerating each growing season from a perennial basal system.

Persistent (compare deciduous). Remaining attached at maturity or at the end of the growing season.

Petiole. The stalk of a leaf; in ferns called a stipe.

Phloem. Vascular tissue that conducts soluble organics and hormones; located outside the xylem.

Pilose. Shaggy with soft, long, straight hairs.

Pinna (pl. pinnae). A leaflet of a pinnate leaf; the first division of a compound leaf.

Pinnate. Once divided with the divisions or segments separately attached to the rachis (fig. 3).

Pinnate-pinnatifid. A pinnate leaf with the leaflets divided from about one quarter to more than halfway to the midvein; a frond that is not quite bipinnate (fig. 3).

Pinnatifid. Divided into narrow lobes extending from about one quarter to more than halfway to the midvein or rachis (fig. 3).

Pinnule. A secondary pinna (leaflet); the ultimate segment of a frond that is at least bipinnate or more compound.

Poikilohydrous. Said of a plant adapted to dry periods, when leaves shrivel and curl, but they can open and refreshen soon after moisture returns.

Polymorphic. Having many forms.

Polyploid. A condition in which there are at least three sets of chromosomes per nucleus; triploid, tetraploid, etc.

Prickly. A surface with small, weak, spinelike projections.

Prostrate. Lying flat on the ground.

Prothallus (pl. prothalli). The vegetative gametophyte stage of pteridophytes, resulting from spore germination, usually dorsiventral and cordate, bearing antheridia and archegonia; also prothallium.

Proximal (compare distal). Toward the base.

Pteridophyte. The collective name for ferns and fern allies.

Pubescent (compare glabrous). Bearing any kind of hairs, covered with short, soft hairs.

Pustulate. With tiny blisters.

Putative. Supposed or probable.

Quadripinnate. Four times pinnate.

Rachis. The main axis of a frond or compound leaf.

Rain forest. A woodland with rainfall of at least 100 inches a year, characterized by dense and complex vegetation, plants usually broadleaved and evergreen, with intermingling canopies and undergrowth.

Recurved. Curved downward.

Reflexed. Bent downward or backward.

Reniform. Kidney-shaped.

Reticulate. A network of veins.

Rhizoid. A hairlike or filamentous structure on a gametophyte, usually anchoring the gametophyte thallus.

Rhizome. Modified stem at or under ground level, from which adventitious roots and fronds are produced.

Rhizophore. A specialized leafless stem that bears roots, as described in some *Selaginella* species.

Rhombic. Diamond-shaped.

Rhomboid. Nearly diamond-shaped with obtuse lateral angles.

Rosette. An arrangement of leaves radiating from a center and usually at or close to the ground.

Rudimentary. Imperfectly developed and not functional.

Rushlike. Resembling rushes (*Juncus, Scirpus, Typha,* etc.).

Scale. A thin, flat scarious structure; a small, dry, usually appressed leaf, vestigial leaf, or bract.

Segment. Each free part of a divided frond.

Septate. Partitioned, multicellular.

Serrate. With forward-projecting sharp teeth.

Serrulate. Finely serrate.

Sessile. Attached directly without a stalk.

Simple. One structure, undivided.

Sorus (pl. sori). A cluster or arrangement of sporangia in ferns.

Spike. In pteridophytes, an axis with sessile sporangia or scales.

Spinulose. With small spines.

Sporangium (pl. sporangia). A spore case.

Spore. In pteridophytes, a single detached cell with a protective, often ornamental wall, with usually haploid nucleus and cytoplasm, capable of developing into a gametophyte, or a new individual.

Sporeling. A juvenile sporophyte, usually still attached to a prothallus.

Sporocarp. A thick-walled organ containing sporangia in heterosporous ferns, as in *Marsilea*.

Sporophore (compare trophophore). A fertile frond; a spore-bearing spike or panicle as in *Ophioglossum*.

Sporophyll. A leaflike structure that bears or subtends sporangia and spores.

Sporophyte. The diploid generation with a full complement of chromosomes; the fern plant that produces spores.

Sporulation. The formation of spores.

Stellate. Starlike; hairs that have radiating branches.

Sterile (compare fertile). Said of plants not producing viable spores or seeds.

Stipe. The petiole of a fern frond; a stalk.

Strobilus (pl. strobili). A cone; an axis from which closely spaced spore-bearing appendages arise, as in *Selaginella* spp.

Subtend. To be below and close to.

Superior. Topmost, positioned above another organ.

Supra. Prefix meaning above.

Supramedial. Above the middle.

Synonymous. Having the same meaning; more than one name referring to a plant taxon.

Taxon (pl. taxa). A taxonomic group at any rank (subspecies, species, genus, family, etc.).

Taxonomy. The study of classification of life forms.

Terminal. At the tip or apex.

Ternate. Palmate with three leaflets.

Terrestrial. Growing in ground.

Tetraploid. Having four sets of chromosomes in each nucleus.

Thallus. An undifferentiated plant body (no roots, stems, or leaves).

Tomentose. With matted woolly hairs.

Tortuous. Twisted.

Translucent. Nearly transparent.

Transverse (compare longitudinal). At right angles to the long or longitudinal axis.

Trifoliate. With three leaves.

Trifoliolate. With three leaflets.

Tripinnate. Three times pinnate (fig. 3).

Trophophore (compare sporophore). A sterile frond that does not bear sporangia.

Truncate. The end straight across, as if cut off.

Tuber. A swollen, subterranean organ.

Tufted. Clumped.

Ultimate segment. The division of a pinna farthest from the costa or rachis.

Umbo. A projecting knob or protuberance.

Vascular. Pertaining to conducting tissues—xylem and phloem—in leaves, stems, or roots.

Vascular bundle. A bundle of vascular tissue; a vein.

Venation. The arrangement of veins.

Ventral (compare dorsal). Pertaining to the front of an organ or the upper surface of a leaf.

Vernation. The arrangement of a leaf or frond in budlike, unexpanded, young condition.

Villous. With long, soft, shaggy but unmatted hairs.

Viscid. Sticky or gummy.

Xeric. Drought-resistant; able to thrive under arid conditions.

Xylem. Vascular tissue that conducts water and dissolved minerals.

Zygote. The fertilized egg.

Literature Cited

Allen, D. E. 1969. *The Victorian Fern Craze: A History of Pteridomania*. London: Hutchinson.

Benham, D. M. 1992. Additional taxa in *Astrolepis*. *American Fern Journal* 82:59–62.

Benham, D. M., and M. D. Windham. 1992. Generic affinities of the star-scaled cloak ferns. *American Fern Journal* 82:47–58.

———. 1993. *Astrolepis*. Pp. 140–3 in *Flora of North America*, Vol. 2: *Pteridophytes and Gymnosperms*. New York: Oxford University Press.

Bold, H. C., C. J. Alexopoulos, and T. Delevoryas. 1987. *Morphology of Plants and Fungi*. 5th ed. New York: Harper and Row.

Correll, D. S. 1955. Pteridophyta. Pp. 1–121 and 39 plates in *Flora of Texas*, C. L. Lundell and Collaborators. Dallas: University Press.

Correll, D. S., and M. C. Johnston. 1970. *Manual of the Vascular Plants of Texas*. Renner: Texas Research Foundation.

Cranfill, R. B. 1993. Dennstaedtiaceae. Pp. 198–205 in *Flora of North America*, Vol. 2: *Pteridophytes and Gymnosperms*. New York: Oxford University Press.

Felger, R. S. 2000. Pteridophytes. Pp. 43–8 in *Flora of the Gran Desierto and Rio Colorado of Northwestern Mexico*. Tucson: University of Arizona Press.

Flora of North America. 1993. Vol. 2: *Pteridophytes and Gymnosperms*. New York: Oxford University Press.

Gastony, G. J. 1986. Electrophoretic evidence for the origin of fern species by unreduced spores. *American Journal of Botany* 73:1563–9.

Gastony, G. J., and M. D. Windham. 1989. Species concepts in pteridophytes: The treatment and definition of agamospermous species. *American Fern Journal* 79:65–77.

Gehlbach, F. R. 1981. *Mountain Islands and Desert Seas*. College Station: Texas A&M University Press.

Harrington, H. D. 1967. *Edible Plants of the Rocky Mountains*. Albuquerque: University of New Mexico Press.

Henrickson, J., and M. C. Johnston. 1986. Vegetation and community types of the Chihuahuan Desert. Pp. 20–39 in *Second Symposium on Resources of the Chihuahuan Desert Region, U.S. and Mexico*, ed. J. C. Barlow, A. M. Powell, and B. N. Timmermann. Alpine, Tex.: Chihuahuan Desert Research Institute.

Johnson, D. M. 1986. Systematics of the New World species of *Marsilea* (Marsileaceae). *Syst. Bot. Monogr.* 11:1–87.

Johnston, M. C. 1990. The vascular plants of Texas: A list, updating the *Manual of the Vascular Plants of Texas*. 2nd ed. Austin, Tex.: M. C. Johnston.

Jones, D. L. 1987. *Encyclopedia of Ferns*. Portland, Ore.: Timber Press.

Kenrick, P., and P. R. Crane. 1997. The origin and early evolution of plants on land. *Nature* 389:33–9.

Lumpkin, T. A. 1993. Azollaceae. Pp. 338–42 in Flora of North America, Vol. 2: *Pteridophytes and Gymnosperms*. New York: Oxford University Press.

Mickel, J. T. 1979. *How to Know the Ferns and Fern Allies*. Dubuque, Iowa: William C. Brown Company, Publishers.

————. 1992. Gymnosperms and Pteri-
dophytes. Pp. 120–431 in *Flora
Novo-Galiciana*, vol. 17, R.
McVaugh. Ann Arbor: University of
Michigan Herbarium.

————. 1994. *Ferns for American Gardens*. New York: Macmillan.

National Park Service. 1995. The vascular flora of Big Bend National Park,
Texas. Unpublished report, printed
from a database, Big Bend National
Park, Tex.

Powell, A. M. 1994. *Grasses of the
Trans-Pecos and Adjacent Areas*. Austin: University of Texas Press.

————. 1998. *Trees and Shrubs of the
Trans-Pecos and Adjacent Areas*. Austin: University of Texas Press.

Raven, P. H., R. F. Evert, and S. E.
Eichhorn. 1999. *Biology of Plants*.
6th ed. New York: W. H. Freeman
and Company, Worth Publishers.

Schmidly, D. J. 1977. *The Mammals of
Trans-Pecos Texas*. College Station:
Texas A&M University Press.

Stace, C. A. 2000. Cytology and
cytogenetics as a fundamental taxonomic resource for the 20th and 21st
centuries. *Taxon* 49:451–77.

Tryon, R. M. 1955. *Selaginella rupestris*
and its allies. *Annals of the Missouri
Botanical Garden* 42:1–99.

Tryon, R. M., and A. F. Tryon. 1982.
Ferns and Allied Plants. New York:
Springer- Verlag.

Windham, M. D. 1987. *Argyrochosma*, a
new genus of cheilanthoid ferns. *American Fern Journal* 77:37–44.

————. 1993a. Pteridaceae. Pp. 122–86
in *Flora of North America*, vol. 2:
Pteridophytes and Gymnosperms. New
York: Oxford University Press.

————. 1993b. New taxa and nomencla-
tural changes in the North American
fern flora. *Contr. Univ. Michigan
Herb.* 19:31–61.

Windham, M. D., and E. W. Rabe.
1993. *Cheilanthes. Pp. 152–69 in
Flora of North America*, vol. 2: *Pteridophytes and Gymnosperms*. New
York: Oxford University Press.

Worthington, R. D. 1995. *Biota of the
Franklin Mountains*. Part II: Flora. El
Paso, Tex.: Floristic Inventories of the
Southwest Program.

Zech, J. C., P. R. Manning, and W. H.
Wagner, Jr. 1998. A new ad-
der's-tongue (*Ophioglossum*:
Ophioglossaceae) for North America.
Sida 18:307–13.

Index

Page numbers of main taxa treatments are in
boldface. Page numbers of illustrations are in
italics.

abbreviations, xii
Acrostichum, 43
   sinuatum, 45
adder's tongue, 33–4
   Chihuahuan Desert, 34
   family, 18, 32–5
   old world, 34
   stalked, 34
Adiantum, 7, 10, 18, 38, **39–40,** 104
   capillus-veneris, 5, 18, **39–40,** *40,* 79,
      82
   tricholepis, 40
Anabaeba azollae, 5, 99
ancestry/evolutionary history of ferns, 1–2
   Equisetum, 29
Anemia, 17, 18, **35–6**
   mexicana, 18, **36,** *37*
Anemiaceae, 15, 18, **35–6**
aota, 31
Argyrochosma, 17, 19, 38, 39, **40–2,** 49,
      62, 68, 104
   limitanea ssp. mexicana, 17, 19, **41,** *42*
   microphylla, 5, 17, 19, 41, **42,** *42*
Aspleniaceae, 16, 20, **80–3**
Asplenium, 6, 7, 10, 17, 20, **80–3**
   palmeri, 20, 80, **81,** 82, *82*
   resiliens, 20, 80, **81,** 82, *82*
   septentrionale, 17, 20, 80, **81,** *83*
   trichomanes, 80
   trichomanes ssp. quadrivalens, 83
   trichomanes ssp. trichomanes, 20,
      81–3, *82*
Astrolepis, 14, 17, 19, 38, **43–7,** 49, 62
   cochisensis, 4, 19, 43, **44,** *45, 46,* 46, 47
   cochisensis ssp. chihuahuensis, 44
   cochisensis ssp. cochisensis, 44
   integerrima, 19, 44, **45,** *46,* 47
   sinuata, 19, 44, **45,** *46,* 46, 47, 52
   windhamii, 19, 44, **45–7,** *46*
Australian mulga fern, 4
Azolla, xv, 5, 17, 21, **99–100**

   caroliniana, 100
   mexicana, 21, **100,** *100*
   pinnata, 100
Azollaceae, 12, 15, 21, **99–100**

bladder fern, 84–6
   bulblet, 85
   Utah, 86
bommer, hairy, 47
Bommeria, 19, 38, **47**
   dancing, 47
   hispida, 17, 19, **47,** *48,* 68
bracken fern, 4, 76–8
   downy, 77–8
   western, 77–8
brittle fern, 84–6
   southwestern, 85
bulblet bladder fern, 85

Calamites, 2
canuela, 31
chalk ledge fern, 36
Cheilanthes, xv, 7, 10, 12, 14, 19, 39, 41,
      43, **47–61,** 62
   aemula, 19, **51–2,** *53,* 61
   alabamensis, 19, 51, **52,** *53,* 61
   aliena, 63
   aschenborniana, 63
   bonariensis, 17, 19, 51, **52,** *54*
   candida var. copelandii, 63
   castanea, 52, 54
   cochisensis, 44
   eatonii, 19, 50, **52–4,** *55,* 60
   eatonii forma castanea, 52, 54
   feei, 19, 51, **54–6,** *55*
   fendleri, 19, 50, 55, **56**
   grayi, 64
   greggii, 66
   horridula, 19, 49, **56,** *57*
   integerrima, 45
   kaulfussii, 19, 51, **56–7,** *58*
   lendigera, 19, 51, **57,** *58*
   leucopoda, 56, 57

(Cheilanthes continued)
   lindheimeri, 19, 50, **58–60**, *59*, 61
   nealleyi, 66
   neglecta, 66
   parvifolia, 42
   sieberi, 4
   sinuata, 45
   sinuata var. *cochisensis*, 44
   standleyi, 67
   tomentosa, 19, 49, 56, *57*, **60**
   villosa, 19, 50, *55*, **60**
   wootonii, 19, 50, *59*, 59, **60–1**
   wrightii, 19, 50, 58, **61**
   yavapensis, 19, 50, *59*, 59, 60, **61**
cinnamon fern, 4
cliff brake, 68–74
   creeping, 71
   heartleaf, 71
   intermediate, 71
   ovate leaf, 71–2
   purple, 69–70
   spiny, 73–4
   ternate, 72–3
   Trans-Pecos, 72–3
   Wright, 74
   zigzag, 71–2
cliff fern, 91–3
   New Mexico, 92
   Phillips, 92
cloak fern, 61–8
   Aschenborn, 63
   Cochise scaly, 44
   Copeland, 63–4
   false, 40–2
   foreign, 63
   gray, 64–5
   Gregg, 66
   hybrid, 45
   Maxon, 66
   Mexican, 63
   Nealley, 66
   scaled, 63
   scaly, 43–7
   Standley, 47, 67–8
   star, 67–8
   starscaled, 43–7
   wavyleaf, 45
   wholeleaf, 45
   Windham, 45–7
clubmoss, 1, 13
copper fern, 47
cordaites, 2

cuplet fern, 75–6
   beaded, 75–6
cyanobacterium, 5, 99–100
Cyrtomium
   auriculatum, 90
   umbonata, 91
Cystopteris, 20, **84–6**
   bulbifera, 20, **85**, 86, *86*
   fragilis, 84
   fragilis var. *tenuifolia*, 85
   reevesiana, 20, **85**, 86
   utahensis, 20, 85, **86**
dancing bommeria, 47

Dennstaedtia, 6, 20, **75–6**
   bipinnata, 75
   globulifera, 17, 20, **75–6**, *76*
   punctiloba, 75
Dennstaedtiaceae, 16, 20, **74–8**
distribution/habitat of Trans-Pecos ferns, xi, xviii,
      13–4
   arid adaptation, xv, 14
   maps, distribution, xii
doradilla, 27
Dryopteridaceae, 16, 20, 78, **83–93**
Dryopteris, 20, 84, **87–8**
   cinnamomea, 20, **87**, 88
   filix-mas, 20, **87–8**, *89*
   fragrans, 5
   normalis, 79
   normalis var. *lindheimeri*, 79

eared holly fern, 90–1
Eaton lip fern, 52–4
economic importance of ferns, 2–5
   cultivation, 2, 3; *Adiantum*, 39; *Anemia*,
      36; *Asplenium*, 83; *Azolla*, 99–100;
      Marsilea, 98; *Ophioglossum*, 34;
      Pellaea, 70; xeriscaping, xi, 3
   food, 4; *Pteridium*, 76–7
   fossil fuels, 2
   medicinal, 2–5; *Adiantum*, 39; *Anemia*, 36;
      Asplenium, 80; *Cheilanthes*, 52;
      Dryopteris, 88; *Equisetum*, 30
   ornamental/decorative uses 2–3; *Selaginella*,
      25
   poisonous, 4–5; *Argyrochosma*, 42;
      Astrolepis, 44; *Equisetum*, 30;
      Pteridium, 77–8
   other uses, 5; *Azolla*, 100; *Equisetum*, 30

elevations, xii
Equisetaceae, 1, 15, 18, **29–32**
Equisetum, xv, 1, 2, 13, 17, 18, **30–2**, 103
   arvense, 4, 30
   xferrissii, 18, 30, **31**
   funstonii, 31
   hyemale ssp. affine, 18, 30, **31**, 32
   hyemale var. *robustum*, 31
   kansanum, 31
   laevigatum,18, 30, **31–2**, 32
   prealtum, 31
   robustum, 31

fairy swords, 58–60
false cloak fern, 40–2
   small leaf, 42
   southwestern, 41
female fern, 78–9
Fendler lip fern, 56
fern
   Australian mulga, 4
   bladder, 84–6
   bracken, 76–8
   brittle, 84–6
   chalk ledge, 36
   cinnamon, 4
   cinnamon wood, 87
   cliff, 91–3
   cloak, 61–8
   copper, 47
   cuplet, 75–6
   false cloak, 40–2
   female, 78–9
   flowercup, 92 –3
   fork, 13
   fragile, 84–6
   hay-scented, 75
   holly, 88–91
   lip, 47–61
   maiden, 78–9
   maidenhair, 5, 39–40
   male, 87–8
   marsh, 20, 78–9
   mosquito, 99–100
   ostrich, 4
   shield, 78–9
   shielded sorus, 94–6
   silver, 40–2
   venushair, 39–40
   whisk, 1, 13
   wood, 83–93

flower of stone, 25
flowercup fern, 92–3
fork fern, 13
fragile fern, 84–6

Gymnopteris hispida, 47

holly fern, 88–91
   Chisos, 91
   eared, 90–1
horsetail, 1, 4, 13, 18, 29–32

Isoetaceae, 1
Isoetes, 1, 2, 12, 13

jimmyfern, 4, 44

Kaulfuss lip fern, 56–7

lip fern, 47–61
   Alabama, 52
   beaded, 57
   Bonaire, 52
   Eaton, 52–4
   Fendler, 56
   glandular, 56–7
   golden, 52
   graceful, 61
   Kaulfuss, 56–7
   prickly, 56
   rival, 51–2
   rough, 56
   slender, 54–6
   Texas, 51–2
   villous, 60
   whitefoot, 58–60
   woolly, 60
   Wooton, 60–1
   Wright, 61
   Yavapai, 61
little ebony spleenwort, 81
Lycophyta, 1
lycophytes, 1, 2
Lycopodiaceae, 1
Lycopodium, 1, 2, 13

maiden fern, 78–9
    Lindheimer, 79
maidenhair fern, 5, 39–40, 82
maidenhair fern family, 18–20, 36–74
male fern, 87–8
maps, Trans–Pecos region, xvii
    individual distribution, xii
marsh fern family, 20, 78–9
Marsileaceae, 12, 15, 21, **96–9**
Marsilea, xv, 17, 21, **96–9**, 105, 107
    fournieri, 98
    macropoda, 21, *97, 97,* **98,** 99
    mollis, 21, *97, 97,* **98,** 99
    mucronata, 98
    tenuifolia, 98
    uncinata, 98
    vestita, 21, *97, 97,* **98–9**
    vestita ssp. *tenuifolia*, 98
Matteuccia struthiopteris, 4
measurements, xii
morphology of ferns, 5–9, *7*
    blade dissection, 6–8, *8*
    fronds (stipe/blade), 6
    roots, 5
    scales/hairs, 7
    stems, 5–6
    venation (veins), 8
mosquito fern, 99–100

Notholaena, xv, 6, 12, 14, 17, 19, 38, 41, 43,
    49, **61–8**, 104
    aliena, 19, **63**, *64*, 65
    aschenborniana, 19, 38, 61, 62, **63**, *64*
    aurea, 52
    candida var. *copelandii*, 63
    cochisensis, 44
    copelandii, 19, 62, **63–4**, *65*, 66
    grayi, 63
    grayi ssp. grayi, 19, **64–5**, *64*
    greggii, 19, 62, **66**, 67
    integerrima, 45
    limitanea ssp./var. *mexicana*, 41
    nealleyi, 19, 62, **66**, 67
    neglecta, 19, 62, **66**, 67
    parvifolia, 42
    schaffneri, 66
    schaffneri var. *nealleyi*, 66
    sinuata, 45
    sinuata var. *cochisensis*, 44
    sinuata var. *integerrima*, 45

standleyi, 17, 19, 47, 62, *64,* 64, 67–8

Ophioglossaceae, 15, 18, **32–4**
Ophioglossum, 5, 17, 18, **33–4**, 107
    engelmannii, 34
    petiolatum, 18, 33, **34**, 35
    polyphyllum, 18, 33, **34**, 35
    reticulatum, 33
Osmunda, 5
    cinnamomea, 4

Pellaea, xv, 19, 38, 41, 49, 61, 62, **68–74**
    alabamensis, 52
    aspera, 56
    atropurpurea, 19, **69–70**, *70*
    cardiomorpha, 71
    cordifolia, 19, 69, **71**, *72*
    intermedia, 19, 69, **71**, *72*
    longimucronata, 73
    microphylla, 42
    mucronata, 5
    ovata, 19, 68, *70,* **71–2**
    sagittata var. *cordata*, 71
    ternifolia, 19, 69, **72–3**, *73,* 74
    ternifolia var. *wrightiana*, 74
    truncata, 20, 69, 73, **73–4**
    wrightiana, 20, 69, 73, *73,* **74**
pepperwort, 96–9
Phanerophlebia, 7, 10, 20, 84, **88–91**
    auriculata, 20, **90–1**, *90*
    umbonata, 20, *90,* 90, **91**
Pilularia, 96
Pleopeltis, 20–1, **94–6**
    polylepis var. erythrolepis, xi, 20,
        95–6, *95*
    riograndensis, 21, *95,* 95, **96**
Polypodiaceae, 16, 20–1, 37, **94–6**
Polypodium, 94
    bulbiferum, 85
    erythrolepis, 95
    filix-mas, 87
    globuliferum, 75
    thyssanolepis var. *riograndensis*, 96
polypody family, 20, 94–6
    red scale, xi, 95–6
    Rio Grande scaly, 96
Psilophyta, 1
psilophytes, 1
Psilotaceae, 1

Psilotum, 1, 12, 13
Pteridaceae, xv, 16, 18–20, **36–74**, 104
Pteridium, 20, 75, **76–8**
   aquilinum, 4
   aquilinum var. pubescens, 20, **77–8**,
      77
Pteridophytes, xv
Pterophyta, 1
Pteris atropurpurea, 69
   ternifolia, 72

quillwort, 1, 13

reproduction of ferns, 9–13, *11*
   asexual, 12; apogamy, 12, 14; apospory, 12
   fern allies, 12–13
   hybridization, 13
   life cycle, general, 9–12, *11;* heterosporous,
      12
   reproductive structures, 9, *11*, 12
resurrection plant, 25
   hairy, 27

Salviniaceae, 12
scouring rush, 1, 13, 30–2
   common, 31
   Ferriss, 31
   smooth, 31–2
Selaginella, xv, 1, 2, 3, 12, 13, 17, 18, **22–9**,
      106, 107
   arizonica, 18, 23, 24, **25**, 27, 28
   coryi, 28
   densa, 28
   densa var. scopulorum, 28
   lepidophylla, 3, 4, 14, 17, 18, 22, 23,
      24, **25**, 27
   mutica, 18, 23, **25–6**, 27, 28, 29
   mutica var. limitanea, 18, **25–6**
   mutica var. mutica, 18, **25–6**
   xneomexicana, 18, 23, 24, **26–7**, 28
   peruviana, 18, 23, 24, 25, 26, **27**, 28
   pilifera, 3, 14, 17, 18, 22, 23, 24, **27**
   rupincola, 18, 23, 24, 25, 26, **27–8**
   scopulorum, 18, 23, 24, **28**
   sheldonii, 27
   underwoodii, 18, 23, 25, 26, 27, **28**
   viridissima, 18, 23, 24, **28–9**
   wrightii, 18, 23, 24, **29**

Selaginellaceae, 1, 15, 18, **22–9**
shield fern, 78–9
shielded sorus fern, 94–6
siempre viva, 25
silver fern, 40–2
   Mexican, 41
   small leaf, 42
small leaf false cloak fern, 42
southwestern false cloak fern, 41
Sphenophyta, 1
spikemoss, 1, 13, 18, 22–9
   Arizona, 25
   blunt, 25–6
   green, 28–9
   ledge, 27–8
   New Mexico, 26–7
   Peruvian, 27
   rockloving, 27–8
   Rocky Mountain, 28
   slender, 28–9
   Underwood, 28
   Wright, 29
spleenwort, 6, 80–3
   blackstem, 81
   forked, 81
   little ebony, 81
   maidenhair, 81–3
   Palmer, 81

Tectaria cinnamomea, 87
Thelypteridaceae, 16, 20, **78–9**
Thelypteris, 6, 20, **78–9**
   kunthii, 79
   normalis, 79
   ovata var. lindheimeri, 17, 20, 79, **79**
Tmesipteris, 1, 13
Trans-Pecos region, xi, xv–xviii
   climate, xvi, xviii
   elevations, xii, xvi
   fern flora, xv
   geography, xv, xviii
   geology, xvi
   map, xvii
   mountain systems, xvi
   soils, xvi
   topography, xvi
   vegetation, xv, xvi, xviii

water clover, 96–9
    family, 21, 96–9
    hooked, 98–9
    large foot, 98
    Mexican, 98
whisk fern, 1, 13
wood fern, 20, 83–93
    cinnamon, 87
    fragrant, 5

Woodsia, 7, 10, 20, 84, **91–3**
    mexicana, 92
    neomexicana, 20, **92**, *93*, 93
    obtusa var. *glandulosa*, 92
    obtusa var. *plummerae*, 92
    phillipsii, 20, **92**, *93*, 93
    Plummer, 92–3
    plummerae, 20, 91, **92–3**, *93*
    pusilla var. *glandulosa*, 92